Murray Gell-Mann and the Physics of Quarks

Classic Texts in the Sciences

Classic Texts in the Sciences offers essential readings for anyone interested in the origin and roots of our present-day culture. Considering the fact that the sciences have significantly shaped our contemporary world view, this series not only provides the original texts but also extensive historical as well as scientific commentary, linking the classical texts to current developments. *Classic Texts in the Sciences* presents classic texts and their authors not only for specialists but for anyone interested in the background and the various facets of our civilization.

More information about this series at http://www.springer.com/series/11828

Harald Fritzsch
Editor

Murray Gell-Mann and the Physics of Quarks

Editor
Harald Fritzsch
Physik-Department
Ludwig-Maximilians-Universität
München, Germany

ISSN 2365-9963 ISSN 2365-9971 (electronic)
Classic Texts in the Sciences
ISBN 978-3-030-06382-5 ISBN 978-3-319-92195-2 (eBook)
https://doi.org/10.1007/978-3-319-92195-2

Softcover re-print of the Hardcover 1st edition 2018

This book is published under the imprint Birkhäuser, www.birkhauser-science.com, by the registered company Springer Nature Switzerland AG.
The registered company address is: Gewerbestrasse 11, 6330 Cham, Switzerland

Contents

Murray Gell-Mann

Harald Fritzsch

Murray Gell-Mann was born in New York on September 15, 1929. His father Arthur Gell-Mann came from the city of Czernowitz, which today is part of the Ukraine. He studied at the University of Vienna. In 1911 he came to New York, where he married Pauline Reichstein.

Arthur Gell-Mann started a language school in Manhattan, which was successful for a few years. When it failed in the Great Depression, he got a position in a bank. Arthur Gell-Mann was very interested in science, especially in astronomy, physics and mathematics.

Murray Gell-Mann grew up in the area west of the central park. He was a gifted child and learned to read and write at the early age of 3. When he was 10 years old, he read "Finnegans Wake" of James Joyce, a book, which should play a specific role later in his life. Together with his brother Ben, who was 9 years older than Murray, he explored New York City, in particular the Central Park and the Bronx Park.

When Gell-Mann was 14 years old, he received a scholarship from Yale University, which allowed him to study physics at Yale. Afterwards he went to the Massachusetts Institute of Technology and worked on his Ph.D. His advisor was Victor Weisskopf. He obtained his Ph.D. in 1951 and went as a post-doc to the Institute for Advanced Study in Princeton. The director of the institute, Prof. Robert Oppenheimer, encouraged Gell-Mann to work on problems in elementary particle physics.

In Princeton he met the English woman Margaret Dow, which worked in an institute in Princeton. In 1955 they got married. They had two children, a daughter Lisa, born in 1956, and a son Nicholas, born in 1963.

In 1952 Gell-Mann joined the Physics Department at the University of Chicago and worked in the research group of Enrico Fermi. Gell-Mann was in particular interested in the

H. Fritzsch (✉)
Physik-Department, Ludwig-Maximilians-Universität Physik-Department, München, Germany
e-mail: fritzsch@mppmu.mpg.de

H. Fritzsch (ed.), *Murray Gell-Mann and the Physics of Quarks*, Classic Texts in the Sciences, https://doi.org/10.1007/978-3-319-92195-2_1

new particles, discovered in the cosmic rays—the new baryons, which were called "hyperons", and the new K-mesons. Nobody understood, why these particles were created easily in nuclear collisions, but decayed rather slowly.

In order to understand the peculiar properties of the new hadrons, Gell-Mann introduced a new quantum number, which he called strangeness. The nucleons were assigned strangeness zero. The newly discovered lambda hyperon had strangeness (−1), likewise the three sigma hyperons. The two Xi-hyperons had strangeness (−2) and the negatively charged K-meson had strangeness (−1).

Gell-Mann assumed that the strangeness quantum number was conserved by the strong and electromagnetic interactions, but violated by the weak interaction. Thus the decays of the strange particles into normal particles without strangeness could only proceed via the weak interaction.

The idea of strangeness explained in a simple way, why the new particles were produced copiously in hadronic collisions, but decayed very slowly. In a collision a new particle with strangeness (−1) could be produced by the strong interaction only together with a particle with strangeness (+1). For example, a negatively charged sigma hyperon could be produced together with a positively charged K-meson. However a positively charged sigma hyperon could not be produced together with a negatively charged K-meson, since both particles have strangeness (−1). Likewise two neutrons could not turn into two neutral lambda hyperons.

In 1954 Murray Gell-Mann and Francis Low worked on the renormalization program of quantum electrodynamics (QED). They introduced a new method, which later was called the "renormalization group" method. Gell-Mann and Low calculated the energy dependence of the renormalized coupling constant.

In quantum electrodynamics the effective coupling constant increases with the energy. This was observed later with the LEP accelerator at CERN. One found that the fine structure constant at 200 GeV is about 1/127, while at low energies it is close to 1/137. The observed increase agreed perfectly with the theoretical prediction. The methods of Gell-Mann and Low were very successful later in the theory of quantum chromodynamics.

In 1955 Gell-Mann obtained an offer from the California Institute of Technology in Pasadena, which was initiated by Richard Feynman. In 1956 he moved to Pasadena. One year later he was promoted to full professor—he became the youngest full professor in the Caltech history. In 1967 Gell-Mann obtained the prestigious Robert Andrews Millikan professorship.

In 1957 Gell-Mann started to work with Richard Feynman on a new theory of the weak interactions. They published 1 year later their paper "Theory of the Fermi Interaction". Feynman and Gell-Mann describe the weak interaction by a universal interaction, given by the product of two currents. Each current is the difference of a vector current and an axial-vector current.

The charged lepton current is a product of a charged lepton field and an antineutrino field. The electrons emitted in a beta-decay are left-handed, the emitted antineutrinos righthanded. This theory of Feynman and Gell-Mann was used later in the gauge theory of the electroweak interactions.

In 1961 Gell-Mann invented a new symmetry to describe the new baryons and mesons, found in the cosmic rays and in various accelerator experiments. He used the unitary group SU(3). At the same time such a symmetry was also considered by Yuval Neeman, who worked at the Israeli embassy in London. The baryons and mesons were placed in octets of the group SU(3). The spin 3/2 baryons were described by a 10 representation, the decuplet.

Only nine particles in the decuplet were known in 1961, the four delta resonances (strangeness 0), the three sigma resonances (strangeness −1) and the two chi resonances (strangeness −2). Gell-Mann predicted the existence and the mass of a negatively charged tenth particle with strangeness −3, which he called the omega minus particle.

This particle is unique in the decuplet, since due to its strangeness (−3) it could only decay by the weak interaction. Thus it would have a relatively long lifetime. It was found in 1964 in Brookhaven Laboratory—it had the mass, which Gell-Mann had predicted. In 1969 Gell-Mann received the Nobel prize for his new symmetry. Gell-Mann is interested linguistics and speaks besides English also Italian, French and Spanish. Part of his lecture at the Nobel ceremony was given in Swedish.

Gell-Mann described the symmetry breaking by a SU(3)-octet. He found a mass formula, which was also found by Susumu Okubo in Japan and is called the Gell-Mann–Okubo mass formula. It describes the mass differences among the baryons and mesons very well.

In 1964 Gell-Mann discussed the triplets of SU(3), which he called "quarks". This name appeared first in the novel of James Joyce "Finnegans Wake": Three quarks for Muster Mark. The quarks were the constituents of the hadrons. George Zweig, a graduate student of Gell-Mann, worked in 1964 at CERN. He also introduced the quarks, which he called "aces". But Zweig published his idea only as a CERN preprint. Gell-Mann published a short letter in the European journal "Physics Letters" with the title: "A Schematic Model of Baryons and Mesons".

Three quarks were the constituents of the baryons and mesons, the up quark "u", the down quark "d" and the strange quark "s". The baryons were bound states of three quarks—for example the proton had the structure (uud), the neutron (ddu). The pion consisted of a quark and an antiquark.

The strange particles contained one, two or three s-quarks, corresponding to the strangeness −1, −2, and −3 respectively. The lambda particle had the structure (uds). The omega minus was a bound state of three strange quarks: (sss). The mesons were bound states of a quark and an antiquark. For example, the positively charged K-meson was a bound state of an up-quark and a strange antiquark.

In his letter Gell-Mann also mentioned a possible field theory, to describe the dynamics of the quarks. This theory was very similar to quantum electrodynamics. The electron was replaced by a quark, the photon was replaced by a vector boson, which Gell-Mann called "gluon". Later this theory was modified and became the theory of quantum chromodynamics, which describes the strong interactions.

The quarks had peculiar properties—they had in particular the electric charges 2/3 and −1/3. Since the observed hadrons had integral charges, the quarks could not be real particles. Either they were just mathematical symbols, or they must be confined inside the hadrons.

In 1968 the quarks were found indirectly in the SLAC experiments. In the deep inelastic electron-proton experiments the electrons were deflected by point-like constituents. Richard Feynman described these objects as “partons”. It turned out, that the partons were the quarks. Thus the quarks were not mathematical symbols, but particles, confined inside the hadrons. In 1971 Gell-Mann and Harald Fritzsch described the results of the experiments at SLAC with the light cone current algebra of currents.

The quark model had serious problems. For example, the omega minus particle, a bound state of three strange quarks, placed symmetrically in an s-wave, violated the Pauli principle, since the wave function of the ground state is symmetric. William Bardeen, Harald Fritzsch and Murray Gell-Mann introduced in 1971 a new quantum number for the quarks, which they called the “color quantum number”. The quarks appeared in three colors: red, green and blue. The tranformations of the three colors were described by the color group SU(3).

The hadrons were considered as color singlets. The simplest color singlets are the bound states of a quark and an antiquark, the mesons, or of three quarks, the baryons. The baryon wave functions is anti-symmetric in the color index. The omega minus particles consisted of three strange quarks, a read, a green and a blue strange quark. The wave function is antisymmetric in the three color indices, thus there is no problem with the Pauli principle.

In 1972 Fritzsch and Gell-Mann introduced a gauge theory for the strong interactions. The color quantum number was considered to be a gauge quantum number, like the electric charge in QED. The color symmetry was considered to be an exact symmetry. The gauge bosons were massless gluons, which transformed as an octet of the color group.

Later they called this theory “Quantum Chromodynamics”, QCD. The theory was discussed at the Rochester conference in 1972 at the Fermi National Accelerator Laboratory. In 1973 Harald Fritzsch, Murray Gell-Mann and Heinrich Leutwyler discussed the advantages of this theory in the letter “Advantages of the Color Octet Gluon Picture”.

In 1979 Gell-Mann, Pierre Ramond and Richard Slansky introduced the seesaw mechanism for the neutrino masses. The very small neutrino masses are then related to the masses of the charged leptons and a very heavy Majorana mass for the righthanded neutrino. After 1980 Gell-Mann got interested string theory. He thought that the superstring theory might lead to a theory of all particles and forces, including the gravitational interaction.

For 23 years Gell-Mann was one of the directors of the MacArthur foundation. In 1984 he was one of the founders of the Santa Fe Institute, an interdisciplinary research institute near Santa Fe. In 1993 Gell-Mann retired from the California Institute of Technology. He moved to Santa Fe, New Mexico and worked in the Santa Fe Institute. In 1994 his popular book “The Quark and the Jaguar” was published.

Gell-Mann lives in a big house south-east of the city of Santa Fe, in the hills before the Sangre-de-Christo mountains.

Isospin and SU(3)-Symmetry

Harald Fritzsch

In 1911 Ernest Rutherford discovered that an atom consists of a small positively charged nucleus, surrounded by a cloud of electrons. Almost all of the mass of an atom is located in the nucleus, with a very small contribution from the electron cloud.

Also the atomic nucleus is a composite system. Inside the nucleus are positively charged particles, the protons. The number of protons is equal to the number of electrons in the cloud. The nucleus of hydrogen is just one proton.

In the nucleus of helium are two protons. But the mass of the helium nucleus is not twice the proton mass, but about four times. Rutherford suggested that the nucleus consists of positive protons and of neutral particles, which he called neutrons. The mass of a neutron should be about equal to the mass of the proton.

Rutherford was right—in 1932 the neutron was discovered. It is unstable and decays into a proton, an electron and a neutrino. This decay is due to the weak interactions. But a neutron inside a nucleus is usually stable. Atomic nuclei are bound states of protons and neutrons. They are bound by the strong nuclear force.

Apart from the electric charge protons and neutrons are very similar. In particular their masses are about the same. They are considered as different states of the same particle, the nucleon, since the strong force does not distinguish between proton and neutron. The mass of the neutron is slightly larger than the proton mass. The neutron decays into a proton with a lifetime of about 14 min.

In 1932 Werner Heisenberg introduced an internal symmetry, the isospin symmetry. It is an exact symmetry of the strong interactions. Protons and neutrons have isospin 1/2. They are described by the nucleon wave function, which has two components. The proton is the upper component, the neutron is the lower component:

H. Fritzsch (✉)
Physik-Department, Ludwig-Maximilians-Universität Physik-Department, München, Germany
e-mail: fritzsch@mppmu.mpg.de

H. Fritzsch (ed.), *Murray Gell-Mann and the Physics of Quarks*, Classic Texts in the Sciences, https://doi.org/10.1007/978-3-319-92195-2_2

$$\psi = \begin{pmatrix} p \\ n \end{pmatrix}.$$

The transformations of the isospin are described by the unitary group SU(2). The three generators of the isospin can be described by the three Pauli matrices:

$$\tau = (\tau_1, \tau_2, \tau_3),$$
$$\tau_1 = \begin{pmatrix} 0 & 1 \\ 1 & 0 \end{pmatrix}, \quad \tau_2 = \begin{pmatrix} 0 & -i \\ i & 0 \end{pmatrix}, \quad \tau_3 = \begin{pmatrix} 1 & 0 \\ 0 & -1 \end{pmatrix},$$
$$I_i = \frac{1}{2}\tau_i.$$

The commutation relations of the generators are:

$$(I_1 I_2 - I_2 I_1) = [I_1, I_2] = iI_3,$$
$$[I_2, I_3] = iI_1,$$
$$[I_3, I_1] = iI_2.$$

The proton and neutron have different isospin projections:

$$p: \quad I_3 = +\frac{1}{2}, \qquad n: \quad I_3 = -\frac{1}{2}.$$

The representations of the isospin group are the singlet, the doublet, the triplet, the quadruplet etc. The doublet representation is the fundamental representation of SU(2). The two nucleons, the proton and the neutron, are described by a doublet representation.

Heisenberg assumed that the isospin symmetry is broken only by electromagnetism, but this leads to a problem. If electromagnetism is turned off, the mass of the proton should be equal to the mass of the neutron. If electromagnetism is turned on, one expects that the proton mass is larger than the neutron mass, due to the electromagnetic self-energy. But this is not the case—the neutron mass is about 1.4 MeV larger than the proton mass. Thus the isospin symmetry is violated even in the absence of electromagnetism. We shall see later, that this effect is related to the masses of the quarks.

Hideki Yukawa predicted in 1935 the existence of mesons as the particles, which generate the strong nuclear force. From the range of the strong nuclear force Yukawa estimated, that the mass of this meson should be about 100 MeV. In 1947 these mesons were discovered, the three pions:

$$\pi = \begin{pmatrix} \pi^+ \\ \pi^0 \\ \pi^- \end{pmatrix}.$$

The pions have a small mass, only 140 MeV. They are unstable particles—the two charged pions decay via the weak interactions into a charged myon and its neutrino, the neutral pion decays electromagnetically into two photons.

The pions are an isospin triplet. Later two neutral mesons were observed, the η-meson with a mass of about 548 MeV and the η′-meson with a mass of about 958 MeV. Both mesons are isospin singlets.

In 1951 the four Δ-resonances were discovered in the experiments at the cyclotron in Chicago. These resonances were created in the scattering of pions and nucleons. In the collisions of positive pions and of protons the Δ-resonance with electric charge (+2) was observed. The Δ-resonances have spin 3/2 and are described by a quadruplet of the isospin:

$$\Delta = \begin{pmatrix} \Delta^{++} \\ \Delta^{+} \\ \Delta^{0} \\ \Delta^{-} \end{pmatrix}.$$

The Δ-resonances decay into a nucleon and a pion.

In the year 1947 the four K-mesons (mass ~ 495 MeV) were discovered:

$$\begin{pmatrix} K^{+} & K^{-} \\ K^{0} & \bar{K}^{0} \end{pmatrix}.$$

The K-mesons are an isospin doublet. They decay due to the weak interactions. For example, the positively charged K-meson can decay into a muon and a neutrino, or into two pions. The nine mesons, the three pions, the four K-mesons, the η-meson and the η′-meson, are pseudo-scalar mesons, i.e. particles without spin.

There exist also vector mesons with spin one, the three ρ-mesons with electric charges (+1, 0 and −1) and a mass of ~ 770 MeV, which are described by an isospin triplet. Furthermore there are two isospin singlets, the ω-meson (mass ~ 782 MeV) and the φ-meson (mass ~ 1020 MeV):

$$\begin{pmatrix} \rho^{+} \\ \rho^{0} \\ \rho^{-} \end{pmatrix}, \quad (\omega, \quad \phi).$$

In addition there are four K*-mesons (mass ~ 892 MeV):

$$\begin{pmatrix} K^{*+} & K^{*-} \\ K^{*0} & \bar{K}^{*0} \end{pmatrix}.$$

Thus there exist nine vector mesons.

From the year 1947 new baryons were observed in the cosmic rays, the neutral Λ-baryon (mass ~ 1116 MeV), the three Σ-baryons (mass ~ 1190 MeV) and the two Ξ-baryons (mass ~ 1320 MeV). The Λ-baryon is an isospin singlet, the three Σ-baryons are an isospin triplet and the two Ξ-baryons an isospin doublet. These particles are called "hyperons".

The new hyperons and the K-mesons were created in pairs by the strong interactions, but they decayed rather slowly, due to the weak interactions. For this reason Murray Gell-Mann introduced in 1953 a new quantum number, the "strangeness S". The nucleons have no strangeness, the Λ-baryon and the Σ-baryons have strangeness (−1) and the Ξ-baryons have strangeness (−2). Two K-mesons have strangeness (−1), their antiparticles have strangeness (+1).

This new quantum number is conserved in the strong interactions. This implies that in strong interactions a strange particle cannot be produced alone. It must be produced together with another strange particle. For example, a Λ-baryon with strangeness (−1) is produced together with a K meson of strangeness (+1). The decay of the strange particles is due to the weak interactions. Thus strangeness disappears in the weak decays—it is not conserved by the weak interactions.

In 1961 Murray Gell-Mann combined the isospin and the strangeness by introducing a new internal symmetry, based on the group SU(3). This group is an extension of the isospin group SU(2). It has eight generators, which obey the commutation relations:

$$\left[T_i, T_j\right] = i f_{ijk} T_k.$$

The structure constants f(ijk) are:

$$f_{123} = 1 \quad f_{147} = f_{246} = f_{257} = f_{345} = \frac{1}{2}$$

$$f_{156} = f_{367} = -\frac{1}{2}$$

$$f_{458} = f_{678} = \frac{\sqrt{3}}{2}$$

The smallest representations of the group SU(3) are the triplet, the sextet, the octet and the decuplet. The triplet representation is the fundamental representation of SU(3).

The observed hadrons are members of specific representations of SU(3). The lowest baryons and mesons are members of octet representations—Gell-Mann described this as the "eightfold way". The two nucleons, the Λ-hyperon, the three Σ-hyperons and the two Ξ-hyperons are described by an octet:

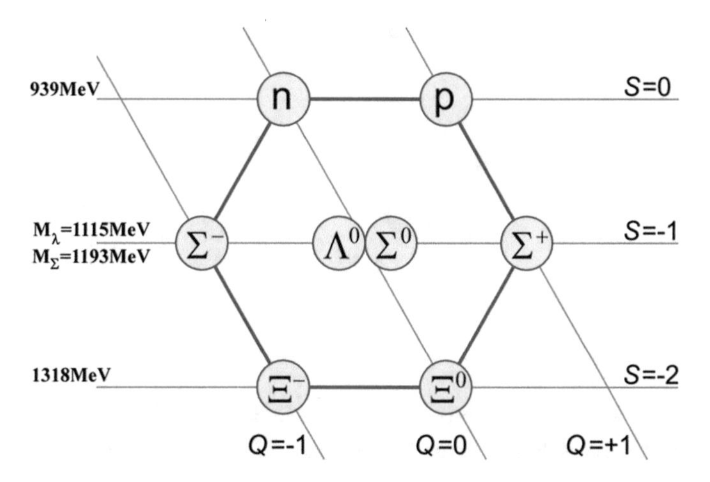

The nine mesons, the three pions, the four K-mesons, the η-meson and the η′-meson, are described by an octet and a singlet:

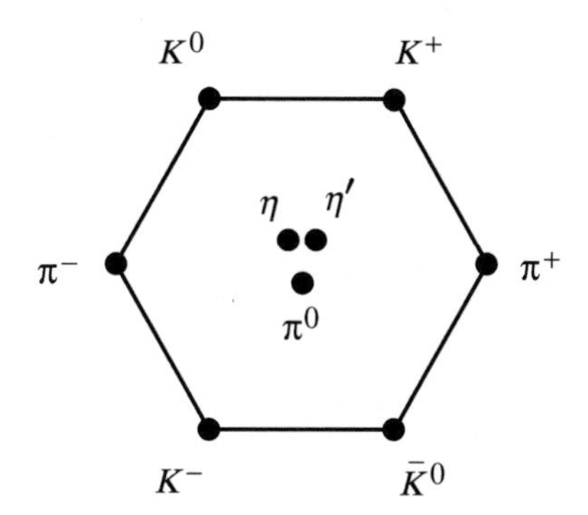

Unlike the isospin symmetry the SU(3) symmetry is strongly broken—the various SU (3) octets have particles with quite different masses. In the baryon octet the masses range from 940 MeV to 1320 MeV. Thus the symmetry breaking is about 25%. When the SU (3) symmetry was introduced, it was unclear, why this symmetry is strongly broken, since there exists no interaction, which breaks the SU(3) symmetry. In the next chapter we shall see that the large symmetry breaking is due to the masses of the quarks.

Besides the four Δ-resonances one observed five excited baryons with spin 3/2: the three Σ-resonances (mass ~ 1380 MeV) and the two Ξ-resonances (mass ~ 1526 MeV). These nine particles could only be described by a decuplet representation of SU(3), but one particle was missing, a baryon with the strangeness (−3). Gell-Mann suggested that this particle, which he called "Ω", must exist.

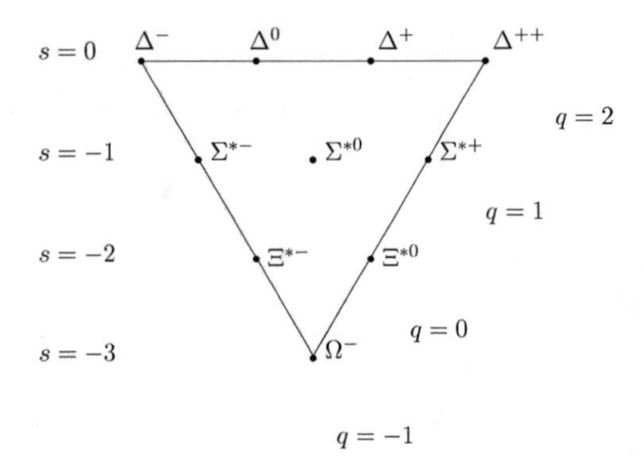

He also predicted the mass of this particle. The Ω-baryon would decay weakly, thus its lifetime would be much longer than the lifetimes of the other members of the baryon decuplet.

The Ω-baryon was discovered in 1964 at the Brookhaven National Laboratory. Its mass is about 1672 MeV. Afterwards it was clear, that the SU(3) symmetry is an approximate symmetry of nature.

The Eightfold Way

Harald Fritzsch

H. Fritzsch (✉)
Physik-Department, Ludwig-Maximilians-Universität Physik-Department, München, Germany
e-mail: fritzsch@mppmu.mpg.de

H. Fritzsch (ed.), *Murray Gell-Mann and the Physics of Quarks*, Classic Texts in the Sciences, https://doi.org/10.1007/978-3-319-92195-2_3

From: California Institute of Technology Laboratory
Report CTSL-20 (1961),

THE EIGHTFOLD WAY: A THEORY OF STRONG INTERACTION SYMMETRY*

Murray Gell-Mann

March 15, 1961
(Second printing: April, 1962)
(Third printing: October, 1963)
(Preliminary version circulated Jan. 20, 1961)

Reprint from *The Eightfold Way*, eds. M. Gell-Mann and Y. Ne'eman (W. A. Benjamin, 1964).

*Research supported in part by the U. S. Atomic Energy Commission Contract No. AT(11-1)-68, and the Alfred P. Sloan Foundation.

2

We attempt once more, as in the global symmetry scheme, to treat the eight known baryons as a supermultiplet, degenerate in the limit of a certain symmetry but split into isotopic spin multiplets by a symmetry-breaking term. Here we do not try to describe the symmetry violation in detail, but we ascribe it phenomenologically to the mass differences themselves, supposing that there is some analogy to the μ-e mass difference.

The symmetry is called unitary symmetry and corresponds to the "unitary group" in three dimensions in the same way that charge independence corresponds to the "unitary group" in two dimensions. The eight infinitesimal generators of the group form a simple Lie algebra, just like the three components of isotopic spin. In this important sense, unitary symmetry is the simplest generalization of charge independence.

The baryons then correspond naturally to an eight-dimensional irreducible representation of the group; when the mass differences are turned on, the familiar multiplets appear. The pion and K meson fit into a similar set of eight particles, along with a predicted pseudoscalar meson χ° having $I = 0$. The pattern of Yukawa couplings of π, K, and χ is then nearly determined, in the limit of unitary symmetry.

The most attractive feature of the scheme is that it permits the description of eight vector mesons by a unified theory of the Yang-Mills type (with a mass term). Like Sakurai, we have a triplet ρ of vector mesons coupled to the isotopic spin current and a singlet vector meson ω° coupled to the hypercharge current. We also have a pair of doublets M and $\bar{M}$, strange vector mesons coupled to strangeness-changing currents that are conserved when the differences are turned off. There is only one coupling constant, in the symmetric limit, for the system of eight vector mesons. There is some experimental evidence for the existence of ω° and M, while ρ is presumably the famous $I = 1$, $J = 1$, π-π resonance.

A ninth vector meson coupled to he baryon current can be accommodated naturally in the scheme.

The most important prediction is the qualitative one that the eight baryons should all have the same spin and parity and that the pseudoscalar and vector mesons should form "octets", with possible additional "singlets".

If the symmetry is not too badly broken in the case of the renormalized coupling constants of the eight vector mesons, then numerous detailed predictions can be made of experimental results.

The mathematics of the unitary group is described by considering three fictitious "leptons", ν, e^-, and μ^-, which may or may not have something to do with real leptons. If there is a connection, then it may throw light on the structure of the weak interactions.

It has seemed likely for many years that the strongly interacting particles, grouped as they are into isotopic multiplets, would show traces of a higher symmetry that is somehow broken. Under the higher symmetry, the eight familiar baryons would be degenerate and form a supermultiplet. As the higher symmetry is broken, the Ξ, Λ, Σ, and N would split apart, leaving inviolate only the conservation of isotopic spin, of strangeness, and of baryons. Of these three, the first is partially broken by electro-magentism and the second is broken by the weak interactions. Only the conservation of baryons and of electric charge are absolute.

An attempt [1, 2] to incorporate these ideas in a concrete model was the scheme of "global symmetry", in which the higher symmetry was valid for the interactions of the π meson, but broken by those of the K. The mass differences of the baryons were thus attributed to the K couplings, the symmetry of which was unspecified, and the strength of which was supposed to be significantly less than that of the π couplings.

The theory of global symmetry has not had great success in predicting experimental results. Also, it has a number of defects. The peculiar distribution of isotopic multiplets among the observed mesons and baryons is left unexplained. The arbitrary K couplings (which are not really particularly weak) bring in several adjustable constants. Furthermore, as admitted in Reference [1] and reemphasized recently by Sakurai [3, 4] in his remarkable articles predicting vector mesons, the global model makes no direct connection between physical couplings and the currents of the conserved symmetry operators.

In place of global symmetry, we introduce here a new model of the higher symmetry of elementary particles which has none of these faults and a number of virtues.

We note that the isotopic spin group is the same as the group of all unitary 2×2 matrices with unit determinant. Each of these matrices can be written as $\exp(iA)$, where A is a hermitian 2×2 matrix. Since there are three independent hermitian 2×2 matrices (say, those of Pauli), there are three components of the isotopic spin.

Our higher symmetry group is the simplest generalization of isotopic spin, namely the group of all unitary 3×3 matrices with unit determinant. There are eight independent traceless 3×3 matrices and consequently the new "unitary spin" has eight components. The first three are just the components of the isotopic spin, the eighth is proportional to the hypercharge Y (which is $+1$ for N and K, -1 for Ξ and $\bar{K}$, 0 for Λ, Σ, π, etc.), and the remaining four are strangeness-changing operators.

Just as isotopic spin possesses a three-dimensional representation (spin 1), so the "unitary spin" group has an eight-dimensional irreducible representation, which we shall call simply **8**. In our theory, the baryons supermultiplet corresponds to this representation. When the symmetry is reduced, then **I** and Y are still conserved but the four other components of unitary spin are

4

not; the supermultiplet then breaks up into Ξ, Σ, Λ, and N. Thus the distribution of multiplets and the nature of strangeness or hypercharge are to some extent explained.

The pseudoscalar mesons are also assigned to the representation **8**. When the symmetry is reduced, they become the multiplets K, $\bar{K}$, π, and χ, where χ is a neutral isotopic singlet meson the existence of which we predict. Whether the PS mesons are regarded as fundamental or as bound states, their Yukawa couplings in the limit of "unitary" symmetry are describable in terms of only two coupling parameters.

The vector mesons are introduced in a very natural way, by an extension of the gauge principle of Yang and Mills [5]. Here too we have a supermultiplet of eight mesons, corresponding to the representation **8**. In the limit of unitary symmetry and with the mass of these vector mesons "turned off", we have a completely gauge-invariant and minimal theory, just like electromagnetism. When the mass is turned on, the gauge invariance is reduced (the gauge function may no longer be space-time dependent) but the conservation of unitary spin remains exact. The sources of the vector mesons are the conserved currents of the eight components of the unitary spin [6].

When the symmetry is reduced, the eight vector mesons break up into a triplet ρ (coupled to the still-conserved isotopic spin current), a singlet ω (coupled to the still-conserved hypercharge current), and a pair of doublets M and $\bar{M}$ (coupled to a strangeness-changing current that is no longer conserved). The particles ρ and ω were both discussed by Sakurai. The ρ meson is presumably identical to the $I = 1$, $J = 1$, π-π resonance postulated by Frazer and Fulco [7] in order to explain the isovector electromagnetic form factors of the nucleon. The ω meson is no doubt the same as the $I = 1$, $J = 0$ particle or 3π resonance predicted by Nambu [8] and later by Chew [9] and others in order to explain the isoscalar form factors of the nucleon. The strange meson M may be the same as the K^* particle observed by Alston *et al* [10].

Thus we predict that the eight baryons have the same spin and parity, that K is pseudoscalar and that χ exists, that ρ and ω exist with the properties assigned to them by Sakurai, and that M exists. But besides these qualitative predictions, there are also the many symmetry rules associated with the unitary spin. All of these are broken, though, by whatever destroys the unitary symmetry, and it is a delicate matter to find ways in which these effects of a broken symmetry can be explored experimentally.

Besides the eight vector mesons coupled to the unitary spin, there can be a ninth, which is invariant under unitary spin and is thus not degenerate with the other eight, even in the limit of unitary symmetry. We call this meson B°. Presumably it exists too and is coupled to the baryon current.

It is the meson predicted by Teller [11] and later by Sakurai [3] and explains most of the hard-core repulsion between nucleons and the attraction between nucleons and anti-nucleons at short distances.

We begin our exposition of the "eightfold way" in the next Section by discussing unitary symmetry using fictitious "leptons" which may have nothing to do with real leptons but help to fix the physical ideas in a rather graphic way. If there is a parallel between these "leptons" and the real ones, that would throw some light on the weak interactions, as discussed briefly in Section VI.

Section III is devoted to the **8** representation and the baryons and Section IV to the pseudoscalar mesons. In Section V we present the theory of the vector mesons.

The physical properties to be expected of the predicted mesons are discussed in Section VII, along with a number of experiments that bear on those properties.

In Section VIII we take up the vexed question of the broken symmetry, how badly it is broken, and how we might succeed in testing it.

II. The "Leptons" as a Model for Unitary Symmetry

For the sake of a simple exposition, we begin our discussion of unitary symmetry with "leptons", although our theory really concerns the baryons and mesons and the strong interactions. The particles we consider here for mathematical purposes do not necessarily have anything to do with real leptons, but there are some suggestive parallels. We consider three leptons, ν, e^-, and μ^-, and their antiparticles. The neutrino is treated on the same footing as the other two, although experience suggests that if it is treated as a four-component Dirac field, only two of the components have physical interaction. (Furthermore, there may exist two neutrinos, one coupled to the electron and the other to the muon.)

As far as we know, the electrical and weak interactions are absolutely symmetrical between e^- and μ^-, which are distinguished, however, from ν. The charged particles e^- and μ^- are separated by the mysterious difference in their masses. We shall not necessarily attribute this difference to any interaction, nor shall we explain it in any way. (If one insists on connecting it to an interaction, one might have to consider a coupling that becomes important only at exceedingly high energies and is, for the time being, only of academic interest). We do, however, guess that the μ-e mass splitting is related to the equally mysterious mechanism that breaks the unitary symmetry of the baryons and mesons and splits the super-multiplets into isotopic multiplets. For practical purposes, we shall put all of these splittings into the mechanical masses of the particles involved.

6

It is well known that in present quantum electrodynamics, no one has succeeded in explaining the e-ν mass difference as an electromagnetic effect. Without prejudice to the question of its physical origin, we shall proceed with out discussion as if that mass difference were "turned on" along with the charge of the electron.

If we now "turn off" the μ-e mass difference, electromagnetism, and the weak interactions we are left with a physically vacuous theory of three exactly similar Dirac particles with no rest mass and no known couplings. This empty model is ideal for our mathematical purposes, however, and is physically motivated by the analogy with the strongly interacting particles, because it is at the corresponding stage of total unitary symmetry that we shall introduce the basic baryon mass and the strong interactions of baryons and mesons.

The symmetric model is, of course, invariant under all unitary transformations on the three states, ν, e^-, and μ^-.

Let us first suppose for simplicity that we had only two particles ν and e^-. We can factor each unitary transformation uniquely into one which multiplies both particles by the same phase factor and one (with determinant unity) which leaves invariant the product of the phase factors of ν and e^-. Invariance under the first kind of transformation corresponds to conservation of leptons ν and e^-. It may be considered separately from invariance under the class of transformations of the second kind (called by mathematicians the unitary unimodular group in two dimensions).

Each transformation of the first kind can be written as a matrix $e^{i\phi}1$, where 1 is the unit 2×2 matrix. The infinitesimal transformation is $1 + i(\delta\phi)1$ and so the unit matrix is the infinitesimal generator of these transformations. The transformations of the second kind are generated in the same way by the three independent traceless 2×2 matrices, which may be taken to be the three Pauli isotopic spin matrices T_1, T_2, T_3. We thus have

$$1 + i \sum_{k=1}^{3} \delta\theta_k \frac{\tau_k}{2} \tag{2.1}$$

as the general infinitesimal transformation of the second kind. Symmetry under all the transformations of the second kind is the same as symmetry under τ_1, τ_2, τ_3, in other words charge independence or isotopic spin symmetry. The whole formalism of isotopic spin theory can then be constructed by considering the transformation properties of this doublet or spinor (ν, e^-) and of more complicated objects that transform like combinations of two or more such leptons.

7

The Pauli matrices τ_k are hermitian and obey the rules

$$\begin{aligned}
\mathrm{Tr}\,\tau_i\,\tau_j &= 2\delta_{ij} \\
[\tau_i,\,\tau_j] &= 2\mathrm{i}\mathrm{e}_{ijk}\,\tau_k \\
\{\tau_i,\,\tau_j\} &= 2\delta_{ij}1
\end{aligned} \tag{2.2}$$

We now generalize the idea of isotopic spin by including the third object μ^-. Again we factor the unitary transformations on the leptons into those which are generated by the 3×3 unit matrix 1 (and which correspond to lepton conservation) and those that are generated by the eight independent traceless 3×3 matrices (and which form the "unitary unimodular group" in three dimensions). We may construct a typical set of eight such matrices by analogy with the 2×2 matrices of Pauli. We call them $\lambda_1 \ldots \lambda_8$ and list them in Table I. They are hermitian and have the properties

$$\begin{aligned}
\mathrm{Tr}\,\lambda_i\,\lambda_j &= 2\delta_{ij} \\
[\lambda_i,\,\lambda_j] &= 2\mathrm{i}\mathrm{f}_{ijk}\,\lambda_k \\
\{\lambda_i,\,\lambda_j\} &= \frac{4}{3}\delta_{ij}1 + 2d_{ijk}\,\lambda_k\ ,
\end{aligned} \tag{2.3}$$

where the f_{ijk} are real and totally antisymmetric like the Kronecker symbols e_{ijk} of (2.2), while the d_{ijk} are real and totally symmetric. These properties follow from the equations

$$\begin{aligned}
\mathrm{Tr}\,\lambda_k\,[\lambda_i,\,\lambda_j] &= 4\mathrm{i}\mathrm{f}_{ijk} \\
\mathrm{Tr}\,\lambda_k\{\lambda_i,\,\lambda_j\} &= 4d_{ijk}
\end{aligned} \tag{2.4}$$

derived from (2.3).

The non-zero elements of f_{ijk} and d_{ijk} are given in Table II for our choice of λ_i. Even and odd permutations of the listed indices correspond to multiplication of f_{ijk} by ± 1 respectively and of d_{ijk} by $+1$.

The general infinitesimal transformation of the second kind is, of course,

$$1 + i\sum_i \delta\theta_1 \frac{\lambda_i}{2} \tag{2.5}$$

by analogy with (2.1). Together with conservation of leptons, invariance under the eight λ_i corresponds to complete "unitary symmetry" of the three leptons.

It will be noticed that λ_1, λ_2, and λ_3 correspond to τ_1, τ_2, and τ_3 for ν and e^- nothing for the muon. Thus, if we ignore symmetry between (ν, e^-) and the muon, we still have conservation of isotopic spin. We also have conservation of λ_8, which commutes with λ_1, λ_2, and λ_3 and is diagonal in our representation. We can diagonalize at most two λ's at the same time and we have chosen them to be λ_3 (the third component of the ordinary isotopic

8

spin) and λ_8, which is like strangeness or hypercharge, since it distinguishes the isotopic single μ^- from the isotopic doublet (ν, e^-) and commutes with the isotopic spin.

Now the turning-on of the muon mass destroys the symmetry under λ_4, λ_5, λ_6, and λ_7 (i.e., under the "strangeness-changing" components of the "unitary spin") and leaves lepton number, "isotopic spin", and "strangeness" conserved. The electromagnetic interactions (along with the electron mass) then break the conservation of λ_1 and λ_2, leaving lepton number λ_3, and strangeness conserved. Finally, the weak interactions allow the strangeness to be changed (in muon decay) but continue to conserve the lepton number n_ℓ and the electric charge

$$Q = \frac{\mathrm{e}}{2}\left(\lambda_3 + \frac{\lambda_8}{\sqrt{3}} - \frac{4}{3}n_\ell\right) \tag{2.6}$$

where n_ℓ is the number of leptons minus the number of antileptons and equals 1 for ν, e^-, and μ^- (i.e., the matrix 1).

We see that the situation is just what is needed for the baryons and mesons. We transfer the symmetry under unitary spin to them and assign them strong couplings and basic symmetrical masses. Then we turn on the splittings, and the symmetry under the 4th, 5th, 6th, and 7th components of the unitary spin is lifted, leaving baryon number, strangeness, and isotopic spin conserved. Electromagnetism destroys the symmetry under the 1st and 2nd components of the spin, and the weak interactions destroy strangeness conservation. Finally, only charge and baryon number are conserved.

III. Mathematical Description of the Baryons

In the cased of isotopic spin **I**, we know that the various possible charge multiplets correspond to "irreducible representations" of the simple 2×2 matrix algebra described above for (ν, e^-). Each multiplet has $2I + 1$ components, where the quantum number I distinguishes one representation from another and tells us the eigenvalue $I(I+1)$ of the operator $\sum_{i=1}^{3} I_i^2$, which commutes with all the elements of the isotopic spin group and in particular with all the elements of the isotopic spin group and in particular with all the infinitesimal group elements $1 + i\sum_{i=1}^{3} \delta\,\theta_i\, I_i$. The operators I_i are represented, within the multiplet, by hermitian $(2I+1) \times (2I+1)$ matrices having the same commutation rules

$$[I_i,\, I_j] = \mathrm{i}\mathrm{e}_{ijk} I_k \tag{3.1}$$

as the 2×2 matrices $\tau_i/2$. For the case of $I = 1/2$, we have just $I_i = \tau_i/2$ within the doublet.

If we start with the doublet representation, we can build up all the others by considering superpositions of particles that transform like the original doublet. Thus, the antiparticles e^+, $-\bar{\nu}$ also form a doublet. (Notice the minus sign on the anti-neutrino state or field). Taking $\frac{e^+e^-+\bar{\nu}\nu}{\sqrt{2}}$, we obtain a singlet, that is, a one-dimensional representation for which all the I_i are zero. Calling the neutrino and electron e_α with $\alpha = 1, 2$, we can describe the singlet by $\frac{1}{\sqrt{2}}\bar{e}_\alpha\, e_\alpha$ or, more concisely, $\frac{1}{\sqrt{2}}\bar{e}e$. The three components of a triplet can be formed by taking

$$e^+\nu = \frac{1}{2}\bar{e}(\tau_1 - i\tau_2)e, \quad \frac{e^+e^- + \bar{\nu}\nu}{\sqrt{2}} = \frac{1}{\sqrt{2}}\bar{e}\tau_3 e,$$

and

$$\nu e^- = \frac{1}{2}\bar{e}(\tau_1 - i\tau_2)e.$$

Rearranging these, we have just $\frac{1}{\sqrt{2}}\bar{e}\tau_j e$ with $j = 1, 2, 3$. Among these three states, the 3×3 matrices I_i^{jk} of the three components of I are given by

$$I_i^{jk} = -\mathrm{i}e_{ijk}\,. \tag{3.2}$$

Now let us generalize these familiar results to the set of three states ν, e^-, and μ^-. Call them ℓ_α with $\alpha = 1, 2, 3$ and use $\bar{\ell}\ell$ to mean $\bar{\ell}_\alpha\,\ell_\alpha$, etc. For this system we define $F_i = \lambda i/2$ with $i = 1, 2, \ldots, 8$, just as $I_i = \tau_i/2$ for isotopic spin. The F_i are the 8 components of the unitary spin operator **F** in this case and we shall use the same notation in all representations. The first three components of **F** are identical with the three components of the isotopic spin **I** in all cases, while F_8 will always be $\frac{\sqrt{3}}{2}$ times the hypercharge Y (linearly related to the strangeness). In all representations, then, the components of **F** will have the same commutation rules

$$[F_\mathrm{i},\, F_\mathrm{j}] = \mathrm{i}f_{ijk}F_k \tag{3.3}$$

that they do in the simple lepton representation for which $F_i = \lambda_i/2$. (Compare the commutation rules in (2.3)). The trace properties and anticommutation properties will not be the same in all representations any more than they are for **I**. We see that the rules (3.1) are just a special case of (3.3) with indices 1, 2, 3, since the f's equal the e's for these values of the indices.

We must call attention at this point to an important difference between unitary or **F** spin and isotopic or **I** spin. Whereas, with a simple change of sign on $\bar{\nu}$, we were able to construct form $\bar{e}_\alpha$ a doublet transforming under **I** just like e_α, we are not able to do the same thing for the **F** spin when we consider the three anti-leptons $\bar{\ell}_\alpha$ compared to the three leptons ℓ_α. True, the anti-leptons do give a representation for **F**, but it is, in mathematical language, *inequivalent* to the lepton representation, even though it also has three dimensions. The reason is easy to see: when we go from leptons to

10

anti-leptons the eigenvalues of the electric charge, the third component of **I**, and the lepton number all change sign, and thus the eigenvalues of F_8 change sign. But they were $\frac{1}{2\sqrt{3}}$, $\frac{1}{2\sqrt{3}}$, and $\frac{-1}{\sqrt{3}}$ for leptons and so they are a different set for anti-leptons and no similarity transformation can change one representation into the other. We shall refer to the lepton representation as 3 and the anti-lepton representation as $\bar{\mathbf{3}}$.

Now let us consider another set of "particles" L_α transforming exactly like the leptons ℓ_α under unitary spin and take their antiparticles $\bar{L}_\alpha$. We follow the same procedure used above for the isotopic spin and the doublet e. We first construct the state $\frac{1}{\sqrt{3}}\bar{L}_\alpha\,\ell_\alpha$ or $\frac{1}{\sqrt{3}}\bar{L}\,\ell$. Just as $\frac{\bar{\mathrm{e}}\mathrm{e}}{\sqrt{2}}$ gave a one-dimensional representation of **I** for which all the I_i were zero, so $\frac{\bar{L}\ell}{\sqrt{3}}$ gives a one-dimensional representation of **F** for which all the F_i are zero. Call this one-dimensional representation **1**.

Now, by analogy with $\frac{\bar{\mathrm{e}}T_i\mathrm{e}}{\sqrt{2}}$ with $i = 1, 2, 3$, we form $\frac{\bar{L}\lambda_i\,\ell}{\sqrt{2}}$ with $i = 1, 2, \ldots, 8$. These states transform under unitary spin **F** like an irreducible representation of dimension 8, which we shall call **8**. In this representation, the 8×8 matrices F_1^{jk} of the eight components F_i of the unitary spin are given by the relation

$$F_i^{jk} = -\mathrm{i}\mathrm{f}_{ijk}\ , \tag{3.4}$$

analogous to (3.2).

When we formed an isotopic triplet from two isotopic doublets, in the discussion preceding (3.2), we had to consider linear combinations of the $\frac{\bar{\mathrm{e}}T_i\mathrm{e}}{\sqrt{2}}$ in order to get simple states with definite electric charges, etc. We must do the same here. Using the symbol $\sim$ for "transforms like", we define

$$\begin{aligned}
\Sigma^+ &\sim \frac{1}{2}\bar{L}(\lambda_1 - i\lambda_2)\ell \sim D_\nu^+ \\
\Sigma^- &\sim \frac{1}{2}\bar{L}(\lambda_1 + i\lambda_2)\ell \sim D^\circ \mathrm{e}^- \\
\Sigma^\circ &\sim \frac{1}{\sqrt{2}}\bar{L}\lambda_3\ell \sim \frac{D_\nu^\circ - D^+\mathrm{e}^-}{\sqrt{2}} \\
p &\sim \frac{1}{2}\bar{L}(\lambda_4 - i\lambda_5)\ell \sim S_\nu^+ \\
n &\sim \frac{1}{2}\bar{L}(\lambda_6 - i\lambda_7)\ell \sim S^+\mathrm{e}^- \\
\Xi^\circ &\sim \frac{1}{2}\bar{L}(\lambda_6 + i\lambda_7)\ell \sim D^+\mu^- \\
\Xi^- &\sim \frac{1}{2}\bar{L}(\lambda_4 + i\lambda_5)\ell \sim D^\circ\mu^-
\end{aligned}$$

11

$$\Lambda \sim \frac{1}{\sqrt{2}} \bar{L} \lambda_8 \ell \sim \frac{(D^\circ_\nu + D^+ \mathrm{e}^- - 2S^+ \mu^-)}{\sqrt{6}} . \tag{3.5}$$

The most graphic description of what we are doing is given in the last column, where we have introduced the notation D°, D^+, and S^+ for the $\bar{L}$ particles analogous to the $\bar{\ell}$ particles $\bar{\nu}, \mathrm{e}^+$, and μ^+ respectively. D stands for doublet and S for singlet with respect to isotopic spins, electric charges, and hypercharges of the multiplets are exactly as we are accustomed to this of them for the baryons listed.

We say, therefore, that the eight known baryons form one degenerate supermultiplet with respect to unitary spin. When we introduce a perturbation that transforms like the μ-e mass difference, the supermultiplet will break up into exactly the known multiplets. (Of course, D will split from S at the same time as e^-, ν from μ^-.)

Of course, another type of baryon is possible, namely a singlet neutral one that transforms like $\frac{1}{\sqrt{3}} \bar{L}\, \ell$. If such a particle exists, it may be very heavy and highly unstable. At the moment, there is no evidence for it.

We shall attach no physical significance to the ℓ and $\bar{L}$ "particles" out of which we have constructed the baryons. The discussion up to this point is really just a mathematical introduction to the properties of unitary spin.

IV. **Pseudoscalar Mesons**

We have supposed that the baryon fields N_j transform like an octet **8** under **F**, so that the matrices of **F** for the baryon fields are given by (3.4). We now demand that all mesons transform under **F** in such a way as to have **F**-invariant strong couplings. If the 8 mesons π_i are to have Yukawa couplings, they must be coupled to $\bar{N}\theta_i N$ for some matrices θ_i, and we must investigate how such bilinear forms transform under **F**.

In mathematical language, what we have done in Section III is to look at the direct product $\bar{\mathbf{3}} \times \mathbf{3}$ of the representations $\bar{\mathbf{3}}$ and **3** and to find that it reduces tot he direct sum of **8** and **1**. We identified **8** with the baryons and, for the time being, dismissed **1**. What we must do now is to look at $\bar{\mathbf{8}} \times \mathbf{8}$. Now it is easy to show that actually $\bar{\mathbf{8}}$ is equivalent to **8**; this is unlike the situation for $\bar{\mathbf{3}}$ and **3**. (We note that the values of Y, I_3, Q, etc., are symmetrically disposed about zero in the **8** representation). So the antibaryons transform essentially like the baryons and we must reduce out the direct product $\mathbf{8} \times \mathbf{8}$. Standard group theory gives the result

$$\mathbf{8} \times \mathbf{8} = \mathbf{1} + \mathbf{8} + \mathbf{8} + \mathbf{10} + \bar{\mathbf{10}} + \mathbf{27} , \tag{4.1}$$

where $\bar{\mathbf{27}} = \mathbf{27}$ (this can happen only when the dimension is the cube of an integer). The representation **27** breaks up, when mass differences are turned on, into an isotopic singlet, triplet, and quintet with $Y = 0$, a doublet and

12

a quartet with $Y = 1$, a doublet and a quartet with $Y = -1$, a triplet with $Y = 2$, and a triplet with $Y = -2$. The representation **10** breaks up, under the same conditions, into a triplet with $Y = 0$, a doublet with $Y = -1$, a quartet with $Y = +1$, and a singlet with $Y = +2$. The conjugate representation $\bar{\mathbf{10}}$ looks the same, of course, but with equal and opposite values of Y. None of these much resembles the pattern of the known mesons.

The **8** representation, occurring twice, looks just the same for mesons as for baryons and is very suggestive of the known π, K, and $\bar{K}$ mesons plus one more neutral pseudoscalar meson with $I = 0$, $Y = 0$, which corresponds to Λ in the baryon case. Let us call this meson χ° and suppose it exists, with a fairly low mass. Then we have identified the known pseudoscalar mesons with a octet under unitary symmetry, just like the baryons. The representations **1**, **10**, $\bar{\mathbf{10}}$ and **27** may also correspond to mesons, even pseudoscalar ones, but presumably they lie higher in mass, some or all of them perhaps so high as to be physically meaningless.

To describe the eight pseudoscalar mesons as belonging to **8**, we put (very much as in (3.5))

$$
\begin{aligned}
\chi^\circ &= \pi_8 \\
\pi^+ &= \frac{(\pi_1 - i\pi_2)}{\sqrt{2}} \\
\pi^- &= \frac{(\pi_1 + i\pi_2)}{\sqrt{2}} \\
\pi^\circ &= \pi_3 \\
K^+ &= \frac{(\pi_4 - i\pi_5)}{\sqrt{2}} \\
K^\circ &= \frac{(\pi_6 - i\pi_7)}{\sqrt{2}} \\
\overline{K^\circ} &= \frac{(\pi_6 + i\pi_7)}{\sqrt{2}} \\
K^- &= \frac{(\pi_4 + i\pi_5)}{\sqrt{2}}
\end{aligned}
\tag{4.2}
$$

and we know then that the matrices of **F** connecting the π_j are just the same as those connecting the N_j, namely $F_i^{jk} = -\mathrm{i}f_{ijk}$.

To couple the 8 mesons invariantly to 8 baryons (say by γ_5), we must have a coupling

$$2i\, g_\circ\, \bar{N}\, \gamma_5\, \theta_i\, N\, \pi_i \tag{4.3}$$

for which the relation

$$[F_i,\, \theta_j] = \mathrm{i}f_{ijk}\, \theta_k \tag{4.4}$$

holds. Now the double occurrence of **8** in (4.1) assures us that there are two independent sets of eight 8×8 matrices θ_i obeying (4.4). One of these sets evidently consists of the F_i themselves. It is not hard to find the other set if we go back to the commutators and anti-commutators of the λ matrices in the **3** representation (2.3). Just as we formed $F_i^{jk} = -\mathrm{i}f_{ijk}$, we define

$$D_i^{jk} = d_{ijk} \tag{4.5}$$

and it is easy to show that the D's also satisfy (4.4). We recall that where the F matrices are imaginary and antisymmetric with respect to the basis we have chosen, the D's are real and symmetric.

Now what is the physical difference between coupling the pseudo-scalar mesons π_1 by means of D_1 and by means of F_1? It lies in the symmetry under the operation

$$\begin{aligned} R: p \leftrightarrow \Xi^-,\ n \leftrightarrow \Xi^\circ,\ \Sigma^+ \leftrightarrow \Sigma^-,\ \Sigma^\circ \leftrightarrow \Sigma^\circ,\ \Lambda \leftrightarrow \Lambda \\ K^+ \leftrightarrow \pm K^-,\ K^\circ \leftrightarrow \pm \overline{K^\circ},\ \pi^+ \leftrightarrow \pm\pi^-, \pi^\circ \leftrightarrow \pm\pi^\circ, \chi^\circ \leftrightarrow \pm\chi^\circ, \end{aligned} \tag{4.6}$$

which is not a member of the unitary group, but a kind of reflection. In the language of N_i, we may say that R changes the sign of the second, fifth, and seventh particles; we note that λ_2, λ_5 and λ_7 are imaginary while the others are real. From Table II we can see that under these sign changes f_{ijk} is odd and d_{ijk} even.

It may be that in the limit of unitary symmetry the coupling of the pseudo-scalar mesons is invariant under R as well as the unitary group. In that case, we choose either the plus sign in (4.6) and the D coupling or else the minus sign and the F coupling. The two possible coupling patterns are listed in Table III.

If only one of the patterns is picked out (case of R-invariance), it is presumably the D coupling, since that gives a large $\Lambda\pi\Sigma$ interaction (while the F coupling gives none) and the $\Lambda\pi\Sigma$ interaction is the best way of explaining the binding of Λ particles in hypernuclei.

In general, we may write the Yukawa coupling (whether fundamental or phenomenological, depending on whether the π_i are elementary or not) in the form

$$L_{\text{int}} = 2ig_\circ \bar{N}\gamma_5 \left[\alpha D_i + (1-\alpha)F_i\right] N\pi_i \,. \tag{4.7}$$

We note that in no case is it possible to make the couplings ΛKN and ΣKN both much smaller than the NπN coupling. Since the evidence from photo-K production seems to indicate smaller effective coupling constants for ΛKN and ΣKN than for NπN (indeed, that was the basis of the global symmetry scheme), we must conclude that our symmetry is fairly badly broken. We shall return to that question in Section VII.

14

A simple way to read off the numerical factors in Table III, as well as those in Table IV for the vector mesons, is to refer to the chart in Table V, which gives the transformation properties of mesons and baryons in terms of the conceptual "leptons" and "L particles" of Section III.

An interesting remark about the baryon mass differences may be added at this point. If we assume that they transform like the μ-e mass difference, that is, like the 8th component of the unitary spin, then there are only two possible mass-difference matrices, F_8 and D_8. That gives rise to a sum rule for baryon masses:

$$\frac{1}{2}(m_N + m_\Xi) = \frac{3}{4}m_\Lambda + \frac{1}{4}m_\Sigma \ , \tag{4.8}$$

which is very well satisfied by the observed masses, much better than the corresponding sum rule for global symmetry.

There is no particular reason to believe, however, that the analogous sum rules for mesons are obeyed.

V. **Vector Mesons**

The possible transformation properties of the vector mesons under **F** are the same as those we have already examined in the pseudoscalar case. Again it seems that for low mass states we can safely ignore the representations **27**, **10**, and $\mathbf{\overline{10}}$. We are left with 1 and the two cases of **8**.

A vector meson transforming according to **1** would have $Q = 0$, $I = 0$, $Y = 0$ and would be coupled to the total baryon current $i\bar{N}\gamma_\mu N$, which is exactly conserved. Such a meson may well exist and be of great importance. The possibility of its existence has been envisaged for a long time.

We recall that the conservation of baryon of baryons is associated with the invariance of the theory under infinitesimal transformations

$$N \to (1 + i\epsilon)N \ , \tag{5.1}$$

where ϵ is a constant. This is gauge-invariance of the first kind. We may, however, consider the possibility that there is also guage invariance of the second kind, as discussed by Yang and Lee [12]. Then we could make ϵ a function of space-time. In the free baryon Lagrangian

$$L_N = -\bar{N}(\gamma_\alpha \partial_\alpha + m_\circ)N \tag{5.2}$$

this would produce a new term

$$L_N \to L_N - i\bar{N}\gamma_\alpha N \partial_\alpha \epsilon \tag{5.3}$$

which can be cancelled only if there exists a neutral vector meson field B_α coupled to the current $\bar{N}\gamma_\alpha N$:

$$L_B = -\frac{1}{4}(\partial_\alpha B_\beta - \partial_\beta B_\alpha)^2$$
$$L_{\text{int}} = \text{if}_\circ \bar{N}\gamma_\alpha N B_\alpha \tag{5.4}$$

and which undergoes the gauge transformation

$$B_\alpha \to B_\alpha + \frac{1}{f_\circ}\partial_\alpha \epsilon \,. \tag{5.5}$$

As Yang and Lee pointed out, such a vector meson is massless and if it existed with any appreciable coupling constant, it would simulate a kind of anti-gravity, for baryons but not leptons, that is contradicted by experiment.

We may, however, take the point of view that there are vector mesons associated with a gauge-invariant Lagrangian plus a mass term, which breaks the gauge invariance of the second kind while leaving inviolate the gauge invariance of the first kind and the conservation law. Such situations have been treated by Glashow [13], Salam and Ward [14], and others, but particularly in this connection by Sakurai [3].

The vector meson transforming according to **1** would then be of such a kind. Teller [11], Sakurai [3], and others have discussed the notion that such a meson may be quite heavy and very strongly coupled, binding baryons and anti-baryons together to make the pseudoscalar mesons according to the compound model of Fermi and Yang [15]. We shall leave this possibility open, but not consider it further here. If it is right, then the Yukawa couplings (4.7) must be treated as phenomenolofical rather than fundamental; from an immediate practical point of view, it may not make difference.

We go on to consider the **8** representation. An octet of vector mesons would break up into an isotopic doublet with $Y = 1$, which we shall call M (by analogy with K – the symbol L is already used to mean π or μ); the corresponding doublet $\bar{M}$ analogous to $\bar{K}$; a triplet ρ with $Y = 0$ analogous to π; and a singlet ω° with $Y = 0$ analogous to χ°

We may tentatively identify M with the K^* reported by Alston *et al* [10] at 884 MeV with a width $\Gamma \approx 15$ MeV for break-up into $\pi + K$. Such a narrow width certainly points to a vector rather than a scalar state. The vector meson ρ may be identified, as Sakurai has proposed, with the $I = 1$, $J = 1$, π-π resonance discussed by Frazer and Fulco [7] in connection with the electromagnetic structure of the nucleon. The existence of ω° has been postulated for similar reasons by Nambu [8], Chew [9], and others.

In principle, we have a choice again between couplings of the **D** and the **F** type for the vector meson octet. But there is no question which is the more reasonable theory. The current $1\bar{N}F_j\gamma_\alpha N$ is the current of the F-spin

16

for baryons and in the limit of unitary symmetry the total F-spin current is exactly conserved. (The conservation of the strangeness-changing currents, those of F_4, F_5, F_6 and F_7, is broken by the mass differences, the conservation of F_2 and F_3 by electromagnetism, and that of F_3 and F_8 separately by the weak interactions. Of course, the current of the electric charge

$$Q = e\left(F_3 + \frac{F_8}{\sqrt{3}}\right) \tag{5.6}$$

is exactly conserved.)

Sakurai has already suggested that ρ is coupled to the isotopic spin current and ω to the hypercharge current. We propose in addition that the strange vector mesons M are coupled to the strangeness-changing components of the F-spin current and that the whole system is completely invariant under $\mathbf{F}$ before the mass-difference have been turned on, so that the three coupling constants (suitably defined) are approximately equal even in the presence of the mass differences.

Now the vector mesons themselves carry F spin and therefore contribute to the current which is their source. The problem of constructing a nonlinear theory of this kind has been completely solved in the case of isotopic spin by Yang and Mills [5] and by Shaw [5]. We have only to generalize their result (for three vector mesons) to the case of F spin and eight vector mesons.

We may remark parenthetically that the Yang-Mills theory is irreducible, in the sense that all the 3 vector mesons are coupled to one another inextricably. We may always make a "reducible" theory by adjoining other, independent vector mesons like the field B_α discussed earlier in connection with the baryon current. It is an interesting mathematical problem to find the set of all irreducible Yang-Mills tricks. Glashow and the author [16] have shown that the problem is the same as that of finding all the simple Lie algebras, one that was solved long ago by the mathematicians. The possible dimensions are 3, 8, 10, 14, 15, 21 and so forth. Our generalization of the Yang-Mills trick is the simplest one possible.

But let us "return to our sheep", in this case the 8 vector mesons. We first construct a completely gauge-invariant theory and then add a mass term for the mesons. Let us call the eight fields $\rho_{i\alpha}$, just as we denoted the eight pseudoscalar fields by π_i. We may think of the N_i, the π_i, and the $\rho_{i\alpha}$ as vectors in an 8-dimensional space. (The index α here refers to the four space-time components of a vector field.) We use our totally antisymmetric tensor f_{ijk} to define a cross product

$$(\mathbf{A} \times \mathbf{B})_i = f_{ijk} A_j B_k \,. \tag{5.7}$$

17

The gauge transformation of the second kind analogous to (5.1) and (5.5) is performed with an eight-component gauge function $\boldsymbol{\phi}$

$$\begin{aligned}
&\mathbf{N} \to \mathbf{N} + \boldsymbol{\phi} \times \mathbf{N} \\
&\boldsymbol{\rho}_\alpha \to \boldsymbol{\rho}_\alpha + \boldsymbol{\phi} \times \boldsymbol{\rho}_\alpha - (2\gamma_\circ)^{-1}\partial_\alpha \boldsymbol{\phi} \\
&\boldsymbol{\pi} \to +\boldsymbol{\pi} + \boldsymbol{\phi} \times \boldsymbol{\pi} \,.
\end{aligned} \tag{5.8}$$

We have included the pseudoscalar meson field for completeness, treating it as elementary. We shall not write the π-N and possible π-π couplings in what follows, since they are not relevant and may simply be added in at the end. The bare coupling parameter is $\gamma_\circ$.

We define gauge-covariant field strengths by the relation

$$\mathbf{G}_{\alpha\beta} = \partial_\alpha \boldsymbol{\rho}_\beta - \partial_\beta \boldsymbol{\rho}_\alpha + 2\gamma_\circ \boldsymbol{\rho}_\alpha \times \boldsymbol{\rho}_\beta \tag{5.9}$$

and the gauge-invariant Lagrangian (to which a common vector meson mass term is presumably added) is simply

$$\begin{aligned}
L = &-\frac{1}{4}\mathbf{G}_{\alpha\beta} \times \mathbf{G}_{\alpha\beta} - m_\circ \bar{\mathbf{N}} \times \mathbf{N} - \bar{\mathbf{N}}\gamma\alpha \times (\partial_\alpha \mathbf{N} + 2\gamma_\circ \boldsymbol{\rho}_\alpha \times \mathbf{N}) \\
&- \frac{1}{2}\mu_\circ^2 \boldsymbol{\pi} \times \boldsymbol{\pi} - \frac{1}{2}(\partial_\alpha \boldsymbol{\pi} + 2\gamma_\circ \boldsymbol{\rho}_\alpha \times \boldsymbol{\pi}) \times (\partial_\alpha \boldsymbol{\pi} + 2\gamma_\circ \boldsymbol{\rho}_\alpha \times \boldsymbol{\pi}) \,.
\end{aligned} \tag{5.10}$$

There are trilinear and quadrilinear interactions amongst the vector mesons, as usual, and also trilinear and quadrilinear couplings with the pseudoscalar mesons. All these, along with the basic coupling of vector mesons to the baryons, are characterized in the limit of no mass differences by the single coupling parameter $\gamma_\circ$. The symmetrical couplings of $\boldsymbol{\rho}_\alpha$ to the bilinear currents of baryons and pseudoscalar mesons are listed in Table IV. In section VII, we shall use them to predict a number of approximate relations among experimental quantities relevant to the vector mesons.

As in the case of the pseudoscalar couplings, the various vector couplings will have somewhat different strengths when the mass differences are included, and some couplings which vanish in (5.10) will appear with small coefficients. Thus, in referring to experimental renormalized coupling constants (evaluated at the physical masses of the vector mesons) we shall use the notation $\gamma_{N\Lambda M}$, $\gamma_{NN\rho}$, etc. In the limit of unitary symmetry, all of these that do not vanish are equal.

VI. **Weak Interactions**

So far, the role of the leptons in unitary symmetry has been purely symbolic. Although we introduced a mathematical F for ν, e^-, and μ^-, that spin is not coupled to the eight vector mesons that take up the F spin gauge

18

for baryons and mesons. If we take it seriously at all, we should probably regard it as a different spin, but one with the same mathematical properties.

Let us make another point, which may seem irrelevant but possibly is not. The photon and the charge operator to which it is coupled have not so far been explicitly included in our scheme. They must be put in as an afterthought, along with the corresponding gauge transformation we have treated. If the weak interactions are carried [17] by vector bosons X_α and generated by a gauge transformation [18, 19] of their own, then these bosons and gauges have been ignored as well. Such considerations might cause us, if we are in a highly speculative frame of mind, to wonder about the possibility that each kind of interaction has its own type gauge and its own set of vector particles and that the algebraic properties of these gauge transformations conflict with one another.

When we draw a parallel between the "F spin" of leptons and the F spin of baryons and mesons, and when we discuss the weak interactions at all, we are exploring phenomena that transcend the scheme we are using. Everything we say in this section must be regarded as highly tentative and useful only in laying the groundwork for a possible future theory. The same is true of any physical interpretation of the mathematics in Sections II and III.

We shall restrict our discussion to charge-exchange weak currents and then only to the vector part. A complete discussion of the axial vector weak currents may involve more complicated concepts and even new mesons [20] (scalar and/or axial vector) lying very high in energy.

The vector weak current of the leptons is just $\bar{\nu}\gamma_\alpha e + \bar{\nu}\gamma_\alpha \mu$. If we look at the abstract scheme for the baryons in (3.5), we see that a baryon current with the same transformation properties under F would consist of two parts: one, analogous to $\bar{\nu}\gamma_\alpha e$, would have $|\Delta\mathbf{I}| = 1$ and $\Delta S = 0$, while the other, analogous to $\bar{\nu}\gamma_\alpha \mu$, would have $|\Delta\mathbf{I}| = 1/2$ and $\Delta S/\Delta Q = +1$. These properties are exactly the ones we are accustomed to associate with the weak interactions of baryons and mesons.

Now the same kind of current we have taken for the leptons can be assigned to the conceptual bosons L of Section III. Suppose it to be of the same strength. Then, depending on the relative sign of the lepton and L weak currents, the matrices in the baryon system may be F's or D's.

Suppose, in the $\Delta S = 0$ case, the relative sign is such as to give F. Then the resulting current is just one component of the isotopic spin current; and the same result will hold for mesons. Thus we will have the conserved vector current that has been proposed [17] to explain the lack of renormalization of the Fermi constant.

In the $\Delta S = 1$ case, by taking the same sign, we could get the almost-conserved strangeness-changing vector current, the current of $F_4 + iF_5$.

Further speculations along these lines might lead to a theory of the weak interactions [21].

VII. **Properties of the New Mesons**

The theory we have sketched is fairly solid only in the realm of the strong interactions, and we shall restrict our discussion of predictions to the interactions among baryons and mesons.

We predict the existence of 8 baryons with equal spin and parity following the pattern of N, Λ, Σ, and Ξ. Likewise, given the π and its coupling constant, we predict a pseudoscalar K and a new particle, the χ°, both coupled (in the absence of mass differences) as in (4.7), and we predict pion couplings to hyperons as in the same equation.

Now in the limit of unitary symmetry an enormous number of selection and intensity rules apply. For example, for the reactions

PS meson $+$ baryon $\to +$ PS meson $+$ baryon,

there are only 7 independent amplitudes. Likewise, baryon-baryon forces are highly symmetric. However, the apparent smallness of $g_1^{2/4\pi}$ for $NK\Lambda$ and $NK\Sigma$ compared to $N\pi N$ indicates that unitary symmetry is badly broken, assuming that it is valid at all. We must thus rely principally on qualitative predictions for tests of the theory; in Section VIII we take up the question of how quantitative testing may be possible.

The most clear-cut new prediction for the pseudoscalar mesons is the existence of χ°, which should decay into 2γ like the π°, unless it is heavy enough to yield $\pi^+ + \pi^- + \gamma$ with appreciable probability. (In the latter case, we must have $(\pi^+\pi^-)$ in an odd state.) $\chi^\circ \to 3\pi$ is forbidden by conservation of I and C. For a sufficiently heavy χ°, the decay $\chi^\circ \to 4\pi$ is possible, but hampered by centrifugal barriers.

Now we turn to the vector mesons, with coupling pattern as given in Table IV. We predict, like Sakurai, the ρ meson, presumably identical with the resonance of Frazer and Fulco, and the ω meson, coupled to the hypercharge. In addition, we predict the strange vector meson M, which may be the same as the K^* of Alston *et al.*

Some of these are unstable with respect to the strong interactions and their physical coupling constants to the decay products are given by the decay widths. Thus, for $M \to K + \pi$, we have

$$\Gamma_M = 2\frac{\gamma^2_{MK\pi}}{4\pi}\frac{k^3}{m^2_M}, \tag{7.1}$$

where k is the momentum of one of the decay mesons. We expect, of course, a $\cos^2\theta$ angular distribution relative to the polarization of M and a charge ratio of $2:1$ in favor of $K^\circ + \pi^+$ or $K^+ + \pi^-$.

20

For the $I = 1$, $J = 1$, π-π response we have the decay $\rho \to 2\pi$ with width

$$\Gamma_\rho = \frac{8}{3} \frac{\gamma_{\rho\pi\pi}^2}{4\pi} \frac{k^3}{m_\rho^2} \,. \tag{7.2}$$

Using a value $m_\rho = 4.5\, m_\pi$, we would $\Gamma \approx m_\pi \frac{\gamma^2}{4\pi}$ and agreement with the theory of Bowcock *et al* [7] would require a value of $\frac{\gamma^2}{4\pi}$ of the order of 2/3. If, now, we assume that the mass of M is really around 880 Mev, the (7.1) yields $\Gamma_M \approx \frac{\gamma^2}{4\pi} \times 50$ MeV. If the width is around 15 Mev, then the two values of $\gamma^2/4\pi$ are certainly of the same order.

We can obtain information about vector coupling constants in several other ways. If we assume, with Sakurai and Dalitz, that the Y^* of Alston *et al* [22] (at 1380 MeV with decay $Y^* \to \pi + \Lambda$) is a bound state of $\bar{K}$ and N in a potential associated with the exchange of ω and ρ, then with simple Schrödinger theory we can roughly estimate the relevant coupling strengths. In the Schrödinger approximation (which is fairly bad, of course) we have the potential

$$V(\text{triplet}) \approx -3\frac{\gamma_{NN\omega}\gamma_{KK\omega}}{4\pi}\frac{e^{-m_\omega r}}{r} + \frac{\gamma_{NN\rho}\gamma_{KK\rho}}{4\pi}\frac{e^{-m_\rho r}}{r} \,. \tag{7.3}$$

If ω has a mass of around 400 MeV (as suggested by the isoscalar form factor of the nucleon), then the right binding results with both $\gamma^2/4\pi$ of the order of 2/3.

A most important result follows if this analysis has any element of truth, since the singlet potential is

$$V(\text{triplet}) \approx -3\frac{\gamma_{NN\omega}\gamma_{KK\omega}}{4\pi}\frac{e^{-m_\omega r}}{r} - 3\frac{\gamma_{NN\rho}\gamma_{KK\rho}}{4\pi}\frac{e^{-m_\rho r}}{r} \,. \tag{7.4}$$

A singlet version of Y^* should exist considerably below the energy of Y^* itself. Call it Y_s^*. If it is bound by more than 100 MeV or so, it is metastable and decays primarily into $\Lambda + \gamma$, since $\Lambda + \pi$ is forbidden by charge independence. Thus, Y_s^* is a fake Σ°, with $I = 0$ and different mass, and may have caused some difficulty in experiments involving the production of Σ° at high energy. If, because of level shifts due to absorption, Y_s^* is not very far below Y^*, then it should be detectable in the same way as Y^*; one should observe its decay into $\pi + \Sigma$.

Bound systems like Y^* and Y_B^* should occur not only for $\bar{K}N$ but also $K\Xi$. (In the limit of unitary symmetry, these come to the same thing.)

The vector coupling constants occur also in several important poles. (For the unstable mesons, these are of course not true poles, unless we perform an analytic continuation of the scattering amplitude onto a second sheet, in which case they become poles at complex energies; they behave almost like true poles, however, when the widths of the vector meson states are small.)

There is the pole at $q^2 = -m_M^2$ in the reactions $\pi^- + p \to \Lambda + K^\circ$ and $\pi^- + p \to \omega + K$; a peaking of K in the forward direction has already been observed in some of these reaction and should show up at high energies in all of them. Likewise, the pole $q^2 = -m_\pi^2$ in the reaction $K + N \to M + N$ should be observable at high energies and its strength can be predicted directly from the width of M. In the reactions $\pi + N \to \Lambda + M$ and $\pi + N \to \Sigma + M$, there is a pole at $q^2 = -m_K^2$ and measurement of its strength can determine the coupling constants $g_{NK\Lambda}^2/4\pi$ and $g_{NK\Sigma}^2/4\pi$ for the K meson.

In πN scattering, we can measure the pole due to exchange of the ρ meson. In KN and $\bar{K}N$ scattering, there are poles from the exchange of ρ and of ω; these can be separated since only the former occurs in the charge-exchange reaction. In NN scattering with charge-exchange, there is a ρ meson pole in addition to the familiar pion pole. Without charge exchange, the situation is terribly complicated, since there are poles from π, ρ, ω, Ξ and B.

When the pole term includes a baryon vertex for the emission or absorption of a vector menson, we must remember that there is a "strong magnetic" term analogous to a Pauli moment as well as the renormalized vector meson coupling constant.

In a relatively short time, we should a considerable body of information about the vector mesons.

VIII. Violations of Unitary Symmetry

We have mentioned that within the unitary scheme there is no way that the coupling constants of K to both $N\Lambda$ and $N\Sigma$ can both be much smaller than 15, except through large violations of the symmetry. Yet experiments on photoproduction of K particles seem to point to such a situation. Even if unitary symmetry exists as an underlying pattern, whatever mechanism is responsible for the mass differences apparently produce a wide spread among the renormalizes coupling constants as well. It is true that the binding of Λ particles in hypernuclei indicates a $\pi\Lambda\Sigma$ coupling of the same order of magnitude as the πNN coupling, but the anomalously small renormalized constants of the K meson indicate that a quantitative check of unitary symmetry will be very difficult.

What about the vector mesons? Let us discuss first the ρ and ω fields, which are coupled to conserved currents. For typical couplings of these fields, we have the relations

$$\gamma_{\rho\pi\pi}^2 = \gamma_\circ^2 z_3(\rho)[V_\pi\rho(0)]^{-2}\,, \tag{8.1}$$

$$\gamma_{\rho NN}^2 = \gamma_\circ^2 z_3(\rho)[V_1\rho(0)]^{-2}\,, \tag{8.2}$$

22

$$\gamma_{\omega NN}^2 = \gamma_\circ^2 z_3(\omega)[V_1\omega(0)]^{-2}\,, \tag{8.3}$$

etc. Here, each renormalized coupling constant is written as a product of the bare constant, a vacuum polarization renormalization factor, and a squared form factor evaluated at zero momentum transfer. The point is that at zero momentum transfer there is no vertex renormalization because the source currents are conserved. To check, for example, the hypothesis that ρ is really coupled to the isotopic spin current, we must check that $\gamma_\circ^2$ in (8.1) is the same as $\gamma_\circ^2$ in (8.2). We can measure (say, by "pole experiments" and by the width of the π-π resonance) the renormalized constants on the left. The quantities V^2 are of the order unity in any case, and their ratio can be measured by studying electromagnetic form factors [23].

The experimental check of "universality" between (8.1) and (8.2) is thus possible, but that tests only the part of the theory already proposed by Sakurai, the coupling of ρ to the isotopic spin current. To test unitary symmetry, we must compare (8.2) and (8.3); but the the ratio $z_3(\rho)/z_3(\omega)$ comes in to plague us. We may hope, of course, that this ratio is sufficiently close to unity to make the agreement striking, but we would like a better way of testing unitary symmetry quantitatively.

When we consider the M meson, the situation is worse, since the source current of M is not conserved in the presence of the mass differences. For each coupling of M, there is a vertex renormalization factor that complicates the comparison of coupling strengths.

An interesting possibility arises if the vector charge-exchange weak current is really given in the $|\Delta S| = 1$ case by the current of $F_4 \pm iF_5$ just as it is thought to be given in the $\Delta S = 0$ case by that of $F_1 \pm iF_2$ (the conserved current) and if the $\Delta S = 0$ and $|\Delta S| = 1$ currents are of equal strength, like the e_ν and μ_ν currents. Then the leptonic $|\Delta S| = 1$ decays show renormalization factors that must be related to the vertex renormalization factors for the M meson, since the source currents are assumed to be the same. The experimental evidence on the decay $K \to \pi +$ leptons then indicates a renormalization factor, in the square of the amplitude, of the order of 1/20. In the decays $\Lambda \to p +$ leptons and $\Sigma^- \to n +$ leptons, both vector and axial vector currents appear to be renormalized by comparable factors.

The width for decay of M into $K + \pi$, if it is really about 15 MeV, indicates that the renormalized coupling constant $\gamma_{K\pi M}^2/4\pi$ is *not* much smaller than $\gamma_{\rho\pi\pi}^2/4\pi \approx 2/3$ and so there is at present no sign of these small factors in the coupling constants of M. It will be interesting, however, to see what the coupling constant $\gamma_{N\Lambda M}^2/4\pi$ comes out, as determined from the pole in $\pi^- + p \to \Lambda + K^\circ$.

We have seen that the prospect is rather gloomy for a quantitative test of unitary symmetry, or indeed of any proposed higher symmetry that is broken by mass differences or strong interactions. The best hope seems to lie in the possibility of direct study of the ratios of bare constants in experiments involving very high energies and momentum transfers, much larger than all masses [24]. However, the theoretical work on this subject is restricted to renormalizable theories. At present, theories of the Yang-Mills type with a mass do not seem to be renormalizable [25], and no one knows how to improve the situation.

It is in any case an important challenge to theoreticians to construct a satisfactory theory of vector mesons. It may be useful to remark that the difficulty in Yang-Mills theories is caused by the mass. It is also the mass which spoils the gauge invariance of the first kind. Likewise, as in the μ-e case, it may be the mass that produces the violation of symmetry. Similarly, the nucleon and pion masses break the conservation of any axial vector current in the theory of weak interactions. It may be that a new approach to the rest masses of elementary particles can solve many of our present theoretical problems.

IX. **Acknowledgments**

The author takes great pleasure in thanking Dr. S. L. Glasshow and Professor R. P. Feynman for their enthusiastic help and encouragement and for numerous ideas, although they bear none of the blame for any errors or defects in the theory. Conversations with Professor R. Block about Lie algebras have been very enlightening.

24

TABLE I. A set of matrices λ_1

$$\lambda_1 = \begin{pmatrix} 0 & 1 & 0 \\ 1 & 0 & 0 \\ 0 & 0 & 0 \end{pmatrix} \quad \lambda_2 = \begin{pmatrix} 0 & -1 & 0 \\ 1 & 0 & 0 \\ 0 & 0 & 0 \end{pmatrix} \quad \lambda_3 = \begin{pmatrix} 1 & 0 & 0 \\ 0 & -1 & 0 \\ 0 & 0 & 0 \end{pmatrix}$$

$$\lambda_4 = \begin{pmatrix} 0 & 0 & 1 \\ 0 & 0 & 0 \\ 1 & 0 & 0 \end{pmatrix} \quad \lambda_5 = \begin{pmatrix} 0 & 0 & -1 \\ 0 & 0 & 0 \\ 1 & 0 & 0 \end{pmatrix} \quad \lambda_6 = \begin{pmatrix} 0 & 0 & 0 \\ 0 & 0 & 1 \\ 0 & 1 & 0 \end{pmatrix}$$

$$\lambda_7 = \begin{pmatrix} 0 & 0 & 0 \\ 0 & 0 & -1 \\ 0 & 1 & 0 \end{pmatrix} \quad \lambda_8 = \begin{pmatrix} \frac{1}{\sqrt{3}} & 0 & 0 \\ 0 & \frac{1}{\sqrt{3}} & 0 \\ 0 & 0 & \frac{-2}{\sqrt{3}} \end{pmatrix}$$

TABLE II. Non-zero elements of f_{ijk} and d_{ijk}. The f_{ijk} are odd under permutations of any two indices while the d_{ijk} are even

ijk	f_{ijk}	ijk	d_{ijk}
123	1	118	$1/\sqrt{3}$
147	1/2	146	$1/2$
156	$-1/2$	157	$1/2$
246	1/2	228	$1/\sqrt{3}$
257	1/2	247	$-1/2$
345	1/2	256	$1/2$
367	$-1/2$	338	$1/\sqrt{3}$
458	$\sqrt{3}/2$	344	$1/2$
678	$\sqrt{3}/2$	355	$1/2$
		366	$-1/2$
		377	$-1/2$
		448	$-1/(2\sqrt{3})$
		558	$-1/(2\sqrt{3})$
		668	$-1/(2\sqrt{3})$
		778	$-1/(2\sqrt{3})$
		888	$-1/\sqrt{3}$

25

TABLE III. Yukawa interactions of pseudoscalar mesons with baryons, assuming pure coupling through D

$$L_{\text{int}}/\text{ig}_\circ = \pi^\circ\{\bar{p}\gamma_5 p - \bar{n}\gamma_5 n + \frac{2}{\sqrt{3}}\overline{\Sigma^\circ}\gamma_5\Lambda + \frac{2}{\sqrt{3}}\bar{\Lambda}\gamma_5\Sigma^\circ - \overline{\Xi^\circ}\gamma_5\Xi^\circ$$

$$+ \overline{\Xi^-}\gamma_5\Xi^-\} + \pi^+\{\sqrt{2}\bar{p}\gamma_5 n + \frac{2}{\sqrt{3}}\overline{\Sigma^+}\gamma_5\Lambda + \frac{2}{\sqrt{3}}\bar{\Lambda}\gamma_5\Sigma^-$$

$$- \sqrt{2}\overline{\Xi^\circ}\gamma_5\Xi^-\} + \text{h.c.}$$

$$+ K^+\{-\frac{1}{\sqrt{3}}\bar{p}\gamma_5\Lambda + \bar{p}\gamma_5\Sigma^\circ + \sqrt{2}\bar{n}\gamma_5\Sigma^- - \frac{1}{\sqrt{3}}\bar{\Lambda}\gamma_5\Xi^- + \overline{\Sigma^\circ}\gamma_5\Xi^-$$

$$+ \sqrt{2}\overline{\Sigma^+}\gamma_5\Xi^\circ\}$$

$$+ \text{h.c.}$$

$$+ K^\circ\{-\frac{1}{\sqrt{3}}\bar{n}\gamma_5\Lambda + \bar{n}\gamma_5\Sigma^\circ + \sqrt{2}\bar{p}\gamma_5\Sigma^+ - \frac{1}{\sqrt{3}}\bar{\Lambda}\gamma_5\Xi^\circ - \overline{\Sigma^\circ}\gamma_5\Xi^\circ$$

$$+ \sqrt{2}\overline{\Sigma^-}\gamma_5\Xi^-\}$$

$$+ \text{h.c.}$$

$$+ \chi^\circ\{-\frac{1}{\sqrt{3}}\bar{p}\gamma_5 p - \frac{1}{\sqrt{3}}\bar{n}\gamma_5 n - \frac{2}{\sqrt{3}}\bar{\Lambda}\gamma_5\Lambda + \frac{2}{\sqrt{3}}\overline{\Sigma^+}\gamma_5\Sigma^+$$

$$+ \frac{2}{\sqrt{3}}\overline{\Sigma^\circ}\gamma_5\Sigma^\circ + \frac{2}{\sqrt{3}}\overline{\Sigma^-}\gamma_5\Sigma^- - \frac{1}{\sqrt{3}}\overline{\Xi^\circ}\gamma_5\Xi^\circ - \frac{1}{\sqrt{3}}\overline{\Xi^-}\gamma_5\Xi^-\}$$

26

TABLE III. (cont.) Yukawa interactions of pseudoscalar mesons with baryons, assuming pure coupling through F

$$
\begin{aligned}
L_{\text{int}}/\text{ig}_{\circ} = {} & \pi^{\circ}(\bar{p}\gamma_5 p - \bar{n}\gamma_5 n + 2\overline{\Sigma^{+}}\gamma_5\Sigma^{+} - 2\overline{\Sigma^{-}}\gamma_5\Sigma^{-} + \overline{\Xi^{\circ}}\gamma_5\Xi^{\circ} - \overline{\Xi^{-}}\gamma_5\Xi^{-}) \\
& + \pi^{+}(\sqrt{2}\bar{p}\gamma_5 n - \sqrt{2}\overline{\Xi^{\circ}}\gamma_5\Xi^{-} - 2\overline{\Sigma^{+}}\gamma_5\Sigma^{\circ} + 2\overline{\Sigma^{\circ}}\gamma_5\Sigma^{-}) \\
& + \text{h.c.} \\
& + K^{+}(-\sqrt{3}\bar{p}\gamma_5\Lambda + \sqrt{3}\bar{\Lambda}\gamma_5\Xi^{-} - \bar{p}\gamma_5\Sigma^{\circ} - \sqrt{2}\bar{n}\gamma\Sigma^{-} + \overline{\Sigma^{\circ}}\gamma_5\Xi^{-} \\
& \quad + \sqrt{2}\overline{\Sigma^{+}}\gamma_5\Xi^{\circ}) \\
& + \text{h.c.} \\
& + K^{\circ}(-\sqrt{3}\bar{n}\gamma_5\Lambda + \sqrt{3}\bar{\Lambda}\gamma_5\Xi^{\circ} - \bar{n}\gamma_5\Sigma^{\circ} - \sqrt{2}\bar{p}\gamma_5\Sigma^{+} - \overline{\Sigma^{\circ}}\gamma_5\Xi^{\circ} \\
& \quad + \sqrt{2}\overline{\Sigma^{-}}\gamma_5\Xi^{-}) \\
& + \text{h.c.} \\
& + \chi^{\circ}(\sqrt{3}\bar{p}\gamma_5 p + \sqrt{3}\bar{n}\gamma_5 n - \sqrt{3}\overline{\Xi^{\circ}}\gamma_5\Xi^{\circ} - \sqrt{3}\overline{\Xi^{-}}\gamma_5\Xi^{-})
\end{aligned}
$$

27

TABLE IV. Trilinear couplings of ρ's to π's and N's

$$
\begin{aligned}
L_{\text{int}}/i\gamma_\circ = {} & M_\alpha^+\{-\sqrt{3}\bar{p}\gamma_\alpha\Lambda + \sqrt{3}\bar{\Lambda}\gamma_\alpha\Xi^- - \bar{p}\gamma_\alpha\Sigma^\circ - \sqrt{2}\bar{n}\gamma_\alpha\Sigma^- \\
& + \overline{\Sigma^\circ}\gamma_\alpha\Xi^- + \sqrt{2}\overline{\Sigma^+}\gamma_\alpha\Xi^\circ - \sqrt{3}K^-\partial_\alpha\chi^\circ + \sqrt{3}\chi^\circ\partial_\alpha K^- \\
& - K^-\partial_\alpha\pi^\circ + \pi^\circ\partial_\alpha K^- - \sqrt{2}\overline{K^\circ}\partial_\alpha\pi^- + \sqrt{2}\pi^-\partial_\alpha\overline{K^\circ}\} \\
& + \text{h.c.} \\
& + M_\alpha^\circ\{-\sqrt{3}\bar{n}\gamma_\alpha\Lambda + \sqrt{3}\bar{\Lambda}\gamma_\alpha\Xi^\circ + \bar{n}\gamma_\alpha\Sigma^\circ - \sqrt{2}\bar{p}\gamma_\alpha\Sigma^+ \\
& - \overline{\Sigma^\circ}\gamma_\alpha\Xi^\circ + \sqrt{2}\overline{\Sigma^-}\gamma_\alpha\Xi^- - \sqrt{3}\overline{K^\circ}\partial_\alpha\chi^\circ + \sqrt{3}\chi^\circ\partial_\alpha K^\circ \\
& + \overline{K^\circ}\partial_\alpha\pi^\circ - \pi^\circ\partial_\alpha\overline{K^\circ} - \sqrt{2}K^-\partial_\alpha\pi^+ + \sqrt{2}\pi^+\partial_\alpha K^-\} \\
& + \text{h.c.} \\
& + \rho_\alpha^+\{\sqrt{2}\bar{p}\gamma_\alpha n - \sqrt{2}\overline{\Xi^\circ}\gamma_\alpha\Xi^- - 2\overline{\Sigma^+}\gamma_\alpha\Sigma^\circ + 2\overline{\Sigma^\circ}\gamma_\alpha\Sigma^- \\
& + \sqrt{2}K^-\partial_\alpha K^\circ - \sqrt{2}K^\circ\partial_\alpha K^- - 2\pi^-\partial_\alpha\pi^\circ + 2\pi^\circ\partial_\alpha\pi^-\} \\
& + \text{h.c.} \\
& + \rho_\alpha^\circ\{\bar{p}\gamma_\alpha p - \bar{n}\gamma_\alpha n + 2\overline{\Sigma^+}\gamma_\alpha\Sigma^+ - 2\overline{\Sigma^-}\gamma_\alpha\Sigma^- + \overline{\Xi^\circ}\gamma_\alpha\Xi^\circ \\
& - \overline{\Xi^-}\gamma_\alpha\Xi^- + K^-\partial_\alpha K^+ - K^+\partial_\alpha K^- - \overline{K^\circ}\partial_\alpha K^\circ \\
& + K^\circ\partial_\alpha\overline{K^\circ} + 2\pi^-\partial_\alpha\pi^+ - 2\pi^+\partial_\alpha\pi^-\} \\
& + \omega_\alpha^\circ\{\sqrt{3}\bar{p}\gamma_\alpha p + \sqrt{3}\bar{n}\gamma_\alpha n - \sqrt{3}\Xi^\circ\gamma_\alpha\Xi^\circ - \sqrt{3}\overline{\Xi^-}\gamma_\alpha\Xi^- \\
& + \sqrt{3}K^-\partial_\alpha K^+ - \sqrt{3}K^+\partial_\alpha K^- + \sqrt{3}\overline{K^\circ}\partial_\alpha K^\circ \\
& - \sqrt{3}K^\circ\partial_\alpha\overline{K^\circ}\}
\end{aligned}
$$

28

TABLE V. Transformation properties of baryons and mesons, assuming pseudoscalar mesons coupled through D

$$K^+ \sim \frac{\mu^+\nu + S^+\overline{D^\circ}}{\sqrt{2}}$$

$$K^\circ \sim \frac{\mu^+\mathrm{e}^- + S^+D^-}{\sqrt{2}}$$

$$\pi^+ \sim \frac{\mathrm{e}^+\nu + D^+\overline{D^\circ}}{\sqrt{2}}$$

$$\pi^\circ \sim \frac{\bar{\nu}\nu - \mathrm{e}^+\mathrm{e}^- + D^\circ\overline{D^\circ} - D^+D^-}{2}$$

$$\pi^- \sim \frac{\bar{\nu}\mathrm{e}^- + D^\circ D^-}{\sqrt{2}}$$

$$\chi^\circ \sim \frac{\bar{\nu}\nu + \mathrm{e}^+\mathrm{e}^- - 2\mu^+\mu^- + D^\circ\overline{D^\circ} + D^+D^- - 2S^+S^-}{\sqrt{12}}$$

$$\overline{K^\circ} \sim \frac{\mathrm{e}^+\mu^- + D^+S^-}{\sqrt{2}}$$

$$K^- \sim \frac{\bar{\nu}\mu^- + D^\circ S^-}{\sqrt{2}}$$

$$p \sim S^+\nu \qquad n \sim S^+\mathrm{e}^-$$

$$\Sigma^+ \sim D^+\nu \qquad \Sigma^\circ \sim \frac{D^\circ\nu - D^+\mathrm{e}^-}{\sqrt{2}}$$

$$\Sigma^- \sim D^\circ\mathrm{e}^- \qquad \Lambda \sim \frac{D^\circ\nu + D^+\mathrm{e}^- - 2S^+\mu^-}{\sqrt{6}}$$

$$\Xi^\circ \sim D^+\mu^- \qquad \Xi^- \sim D^\circ\mu^-$$

$$M^+ \sim \frac{\mu^+\nu - S^+\overline{D^\circ}}{\sqrt{2}}$$

$$M^\circ \sim \frac{\mu^+\mathrm{e}^- - S^+D^-}{\sqrt{2}}$$

29

TABLE V. (Cont.)

$$\rho^+ \sim \frac{e^+\nu - D^+\overline{D^\circ}}{\sqrt{2}}$$

$$\rho^\circ \sim \frac{\bar{\nu}\nu - e^+e^- - D^\circ\overline{D^\circ} + D^+D^-}{2}$$

$$\rho^- \sim \frac{\bar{\nu}e^- - D^\circ D^-}{\sqrt{2}}$$

$$\omega^\circ \sim \frac{\bar{\nu}\nu + e^+e^- - 2\mu^+\mu^- - D^\circ\overline{D^\circ} - D^+D^- + 2S^+S^-}{\sqrt{12}}$$

$$\overline{M^\circ} \sim \frac{e^+\mu^- - D^+S^-}{\sqrt{2}}$$

$$M^- \sim \frac{\bar{\nu}\mu^- - D^\circ S^-}{\sqrt{2}}$$

References

[1] M. Gell-Mann: Phys. Rev. **106**, 1296 (1957).
[2] J. Schwinger: Ann. Phys. **2**, 407 (1957).
[3] J. J. Sakurai: Ann. Phys. **11**, 1 (1960).
[4] J. J. Sakurai: *Vector Theory of Strong Interactions*, unpublished.
[5] C. N. Yang, R. Mills: Phys. Rev. **96**, 191 (1954). Also R. Shaw unpublished.
[6] After the circulation of the preliminary version of this work (January 1961) the author has learned of a similar theory put forward independently and simultaneously by Y. Ne'eman (Nuclear Phys., to be published). Earlier uses of the 3-dimensional unitary group in connection with the Sakata model are reported by Y. Ohnuki at the 1960 Rochester Conference on High Energy Physics. A. Salam and J. Ward (Nuovo Cimento, to be published) have considered related questions. The author would like to thank Dr. Ne'eman and Professor Salam for communicating their results to him.
[7] W. R. Frazer, J. R. Fulco: Phys. Rev. **117**, 1609 (1960). See also J. Bowcock, W. N. Cottingham, D. Lurie: Phys. Rev. Letters **5**, 386 (1960).
[8] Y. Nambu: Phys. Rev. **106**, 1366 (1957).
[9] G. F. Chew: Phys. Rev. Letters **4**, 142 (1960).
[10] M. Alston *et al*: to be published.
[11] E. Teller: Proceedings of the Rochester Conference (1956).
[12] C. N. Yang, T. D. Lee: Phys. Rev. **98**, 1501 (1955).
[13] S. L. Glashow: Nuclear Phys. **10**, 107 (1959).
[14] A. Salam, J. C. Ward: Nuovo Cimento **11**, 568 (1959).
[15] E. Fermi, C. N. Yang: Phys. Rev. **76**, 1739 (1949).
[16] S. L. Glashow, M. Gell-Mann: to be published.
[17] R. P. Feynman, M. Gell-Mann: Phys. Rev. **109**, 193 (1958).
[18] S. Bludman: Nuovo Cimento **9**, 433 (1958).
[19] M. Gell-Mann, M. Levy: Nuovo Cimento **16**, 705 (1960).

30

[20] M. Gell-Mann: Talk at Rochester Conference on High Energy Physics (1960).
[21] Earlier attempts to draw a parallel between leptons and baryons in the weak interactions have been made by A. Gamba, R. E. Marshak, S. Okubo, Proc. Nat. Acad. Sci. **45**, 881 (1959), and Y. Yamahuchi, unpublished. Dr. S. L. Glashow reports that Yamaguchi's scheme has much in common with the one discussed in this paper.
[22] M. Alston, L. W. Alvarez, P. Eberhard, M. L. Good, W. Graziano, H. K. Ticho, S. G. Wojcicki: Phys. Rev. Letters **5**, 518 (1960).
[23] M. Gell-Mann, F. Zachariasen: *Form Factors and Vector Mesons*, to be published.
[24] M. Gell-Mann, F. Zachariasen: *Broken Symmetries and Bare Coupling Constants*, to be published.
[25] Kamefuchi, Umezawa: to be published. Salam, Kumar: to be published.

Introduction to Quark Model

Harald Fritzsch

The fundamental representation of the isospin group SU(2) is the doublet representation. The proton and the neutron are described by a doublet. The fundamental representation of SU(3) is the triplet representation, but in nature there are no hadrons, which transform as a triplet—the hadrons are described only by singlets, octets and decuplets. This feature of the SU(3) symmetry was not understood until 1964.

In this year Murray Gell-Mann suggested that the hadrons are composite particles. The constituents of the hadrons are SU(3) triplets, the "quarks". They are spin (1/2) fermions. There are three quarks, an up quark, a down quark and a strange quark:

$$q => \begin{pmatrix} u \\ d \\ d \end{pmatrix}.$$

The charge of the up quark is (2/3), the charges of the down quark and of the strange quark are $(-1/3)$. The SU(3) transformations are unitary transformations of the three quarks. The isospin subgroup SU(2) is given by the transformations of the u-quark and the d-quark.

Inside the proton are two up quarks and one down quark: p ~ (uud). If the u quark and the d quark are interchanged, one obtains the neutron: n ~ (ddu). The number of strange quarks inside a hadron is related to the strangeness of the hadron—it is minus the number of strange quarks inside the hadron. The Λ-hyperon and the Σ-hyperons have strangeness (-1), e.g. $\Lambda \sim$ (uds). Inside the two Ξ-hyperons are two strange quarks:

$\Xi(0)$~(uss), $\Xi(-)$~(dss).

H. Fritzsch (✉)
Physik-Department, Ludwig-Maximilians-Universität Physik-Department, München, Germany
e-mail: fritzsch@mppmu.mpg.de

H. Fritzsch (ed.), *Murray Gell-Mann and the Physics of Quarks*, Classic Texts in the Sciences, https://doi.org/10.1007/978-3-319-92195-2_4

It would have been better to use a different sign for the strangeness, but this became clear within the quark model, proposed in 1964, 11 years after the introduction of the strangeness.

Here are the eight baryons with their substructure:

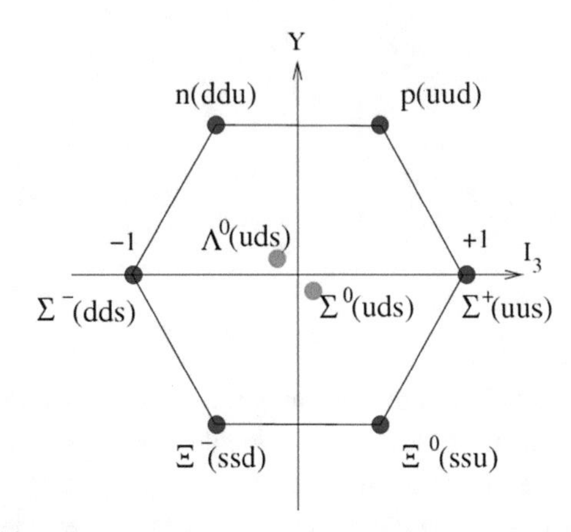

These baryons are fermions with the spin ½. The spin is the sum of the spins of the three quarks. Two quarks are aligned, the spin of the third quark is opposite—thus the sum is ½. There is no contribution from the angular momentum, since the three quarks are in the ground state.

The ten baryon resonances with spin 3/2 have the following substructure:

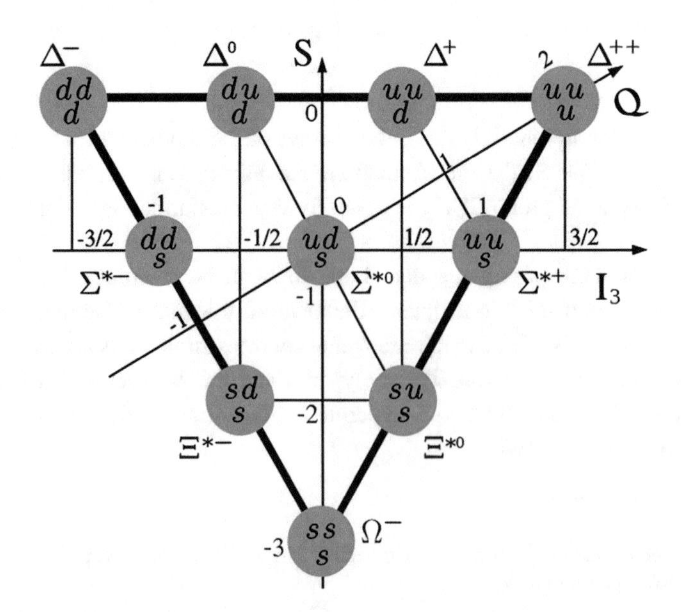

Here the spins of the three quarks are aligned.

The Ω-baryon consists of three strange quarks. It is the only spin (3/2) particle, which does not decay via the strong interactions. It decays weakly and has a relatively long lifetime.

The nine pseudo-scalar mesons are bound states of a quark and an antiquark. The product of a triplet and an anti-triplet gives a singlet and an octet: $3 \times 3^* = 1 + 8$:

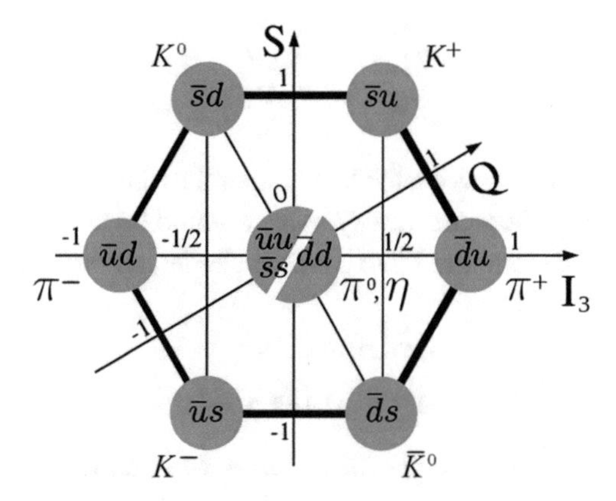

Also the vector mesons are described by an octet and a singlet. The pi mesons are replaced by the rho mesons, the K mesons by the K* mesons, the η-meson by a superposition of the ω-meson and the φ-meson. The ω-meson is mainly a bound state of up and down quarks, the φ-meson is mainly composed of strange quarks. Thus there is a large mixing between the SU(3) octet state and the singlet state:

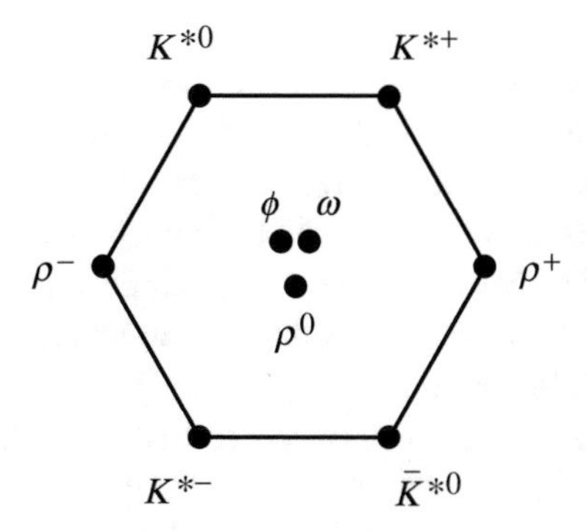

Inside the pseudoscalar mesons the spins of the two quarks are opposite to each other, thus the quarks do not contribute to the angular momentum. Inside the vector mesons the spins are aligned. The spin of a vector meson is provided by the spins of the two quarks.

Gell-Mann thought that the quarks are either mathematical symbols or real particles. In this case they should exist in nature as stable particles with non-integral electric charges.

One has searched for quarks with accelerators, in cosmic rays or in stable matter, but nothing has been found.

Today we think that quarks are real particles, but they do not exist as free particles—they are confined inside the hadrons by the strong force. Nevertheless the quarks have specific masses, which can be deduced from the spectrum of hadrons. Here are the masses of the three quarks:

$$\begin{aligned} m_u &\approx 2.3MeV, \\ m_d &\approx 4.8\,MeV, \\ m_s &\approx 95\,MeV. \end{aligned}$$

The masses of the quarks describe the breaking of the SU(3) symmetry. If the masses of the three quarks would be the same, this symmetry would be unbroken. The mass term for the three quarks can be decomposed into a SU(3)-singlet, an octet and a triplet:

$$\begin{aligned} & m_u\bar{u}u + m_d\bar{d}d + m_s\bar{d}d = \\ & \frac{1}{3}(m_u + m_d + m_s)\cdot\left(\bar{u}u + \bar{d}d + \bar{d}d\right) + \\ & \frac{1}{6}(m_u + m_d - 2m_s)\cdot\left(\bar{u}u + \bar{d}d - 2\bar{s}s\right) + \\ & \frac{1}{2}(m_u - m_d)\cdot\left(\bar{u}u - \bar{d}d\right). \end{aligned}$$

The large difference between the strange quark mass and the up or down quark mass implies the large violation of the SU(3) symmetry. Thus the large symmetry breaking is understood. The isospin symmetry is also broken by the quark mass term—the down quark mass is larger than the up quark mass. This implies that the neutron mass is larger than the proton mass.

Thus far the quark masses cannot be calculated. As the masses of the leptons they are free parameters and have to be measured in the experiments.

In 1974 new heavy hadrons were discovered. They can be described by another quark, the charmed quark. The first new particle, which has been seen, was a vector meson "J/ψ", which is a bound state of a charmed quark and an anti-charmed quark, analogous to the φ-meson—a bound state of a strange quark and an anti-strange quark.

Since the mass of the "J/ψ" meson is about 3.1 GeV, the mass of the charmed quark is quite large, about 1.27 GeV. Hadrons, which consist of a charmed quark, have large masses. The charmed baryon with the substructure "udc" has a mass of about 2286 MeV.

It is useful to describe the four quarks as two doublets:

$$\begin{pmatrix} u \\ d \end{pmatrix} \begin{pmatrix} c \\ s \end{pmatrix}.$$

In 1977 a new heavy meson was discovered at Fermilab, the upsilon meson (Υ), with a mass of about 9.46 GeV. This meson is a bound state of the new quark, the bottom quark b, and its antiquark. The electric charge of the b-quark is ($-1/3$), its effective mass about 4.16 GeV.

B mesons are composed of a bottom antiquark and an up, down, strange or charm quark. The meson, which consists of an anti-up quark and a b quark, has a mass of about 5.28 GeV. The mass of the lightest baryon with the substructure "udb" is about 5620 MeV.

When the b quark was observed, it was assumed that there must also be another quark with the electric charge (2/3), the top quark. In 1995 effects due to the top quark were discovered at the Fermi National Laboratory near Chicago. The mass of the top quark is very large, similar to the mass of a wolfram atom, about 172 GeV.

The top quark decays mainly to a W boson and a bottom quark. The Standard Theory predicts its mean lifetime to be roughly 5×10^{-25} s. This is about a 20th of the timescale for strong interactions—therefore the top quark does not have enough time to form hadrons.

Nature is described by six quarks. Thus far no other quarks have been observed. It is useful to describe the six quarks as three doublets:

$$\begin{pmatrix} u \\ d \end{pmatrix} \begin{pmatrix} c \\ s \end{pmatrix} \begin{pmatrix} t \\ b \end{pmatrix}.$$

The doublets describe the properties of the quarks with respect to the weak interactions. These interactions are generated by the exchange of very massive vector bosons, the weak bosons. There are two charged weak bosons, W(+) and W(−), and a neutral boson, the Z-boson.

If a charged weak boson interacts with a quark, the charge of the quark changes, e.g. u $\Rightarrow$ d. In this way the weak decays are described. For example the beta decay of the neutron follows from the weak transition of a d-quark into a u-quark. The positively charged pion decays into a muon and a neutrino. The quark and the antiquark inside the pion produce a virtual weak boson, which decays into a muon and a neutrino:

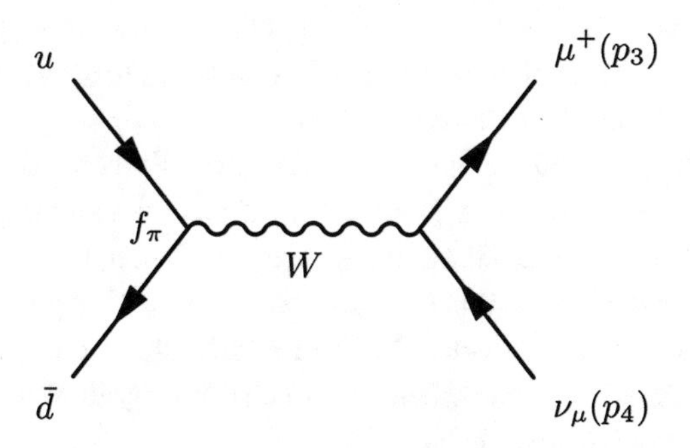

But there is also a weak transition from an s-quark to the u-quark. It is a small effect due to the flavor mixing of the quark. When a weak boson interacts with the u-quark, a superposition of the three quarks d, s and b is produced. This state is described by the three electroweak mixing angles. The transition from the u-quark to the b-quark is very small and we shall neglect it. Then one obtains a mixture of a d-quark and an s-quark:

$$u \Rightarrow d \cdot \cos\theta + s \cdot \sin\theta.$$

The mixing angle is called the "Cabibbo angle". It has been measured to about 13 degrees. Thus far the phenomenon of flavor mixing is still not understood. Presumably the mixing angles are functions of the quark masses.

In 1968 the quarks inside the nucleons were observed. The experiments were carried out at the Stanford Linear Accelerator Center. High energy electrons or positrons were accelerated up to 50 GeV and collided with atomic nuclei. The cross section depends on two variables, the mass of the virtual photon, emitted by the lepton, and the energy transfer from the lepton to the hadron. But it was observed, that at high energies the cross section depends only on the ratio of these two variables.

Richard Feynman pointed out, that this phenomenon, the "scaling behavior" of the cross section, could be understood, if the leptons collided inside the atomic nuclei with point-like constituents. Feynman introduced that name "partons" for these constituents. Later it turned out that the partons were the quarks. Thus the quarks became real particles, which were confined inside the hadrons.

In 1970 Harald Fritzsch and Murray Gell-Mann derived the results of the parton model, using the algebra of currents. In deep inelastic scattering one studies the commutator of two electromagnetic currents near the light cone. Fritzsch and Gell-Mann assumed that this commutator is given by the free quark model. Thus the interaction among the quarks should disappear near the light cone. In this way they could derive the scaling behavior and the results of the parton model.

The quark model has two problems. One is related to the electromagnetic decay of the neutral pion. If the pion is considered to be a bound state of a nucleon and an antinucleon, one can calculate the decay rate. The pion couples to a nucleon-antinucleon pair, which annihilates into two photons. The result agrees with the measured decay rate. But if the pion is assumed to be a bound state of a quark and an antiquark, the decay rate is about an order of magnitude smaller than the observed decay rate.

Another problem is related to the Pauli principle. We consider as an example the Ω-baryon. If quarks are Fermi-Dirac particles, the Pauli principle requires, that the wave function must be anti-symmetric, if two quarks are interchanged. But the wave function of the Ω-baryon is symmetric, since it is a bound state of three strange quarks, and the space wave function is also symmetric, since the three quarks are in the ground state. Thus the Pauli principle is violated. These problems will be solved by introducing a new quantum number, as discussed in the next chapter.

Quarks

Harald Fritzsch

H. Fritzsch (✉)
Physik-Department, Ludwig-Maximilians-Universität Physik-Department, München, Germany
e-mail: fritzsch@mppmu.mpg.de

H. Fritzsch (ed.), *Murray Gell-Mann and the Physics of Quarks*, Classic Texts in the Sciences, https://doi.org/10.1007/978-3-319-92195-2_5

151

Volume 8, number 3 PHYSICS LETTERS 1 February 1964

A SCHEMATIC MODEL OF BARYONS AND MESONS *

M. GELL-MANN
California Institute of Technology, Pasadena, California

Received 4 January 1964

If we assume that the strong interactions of baryons and mesons are correctly described in terms of the broken "eightfold way" [1-3], we are tempted to look for some fundamental explanation of the situation. A highly promised approach is the purely dynamical "bootstrap" model for all the strongly interacting particles within which one may try to derive isotopic spin and strangeness conservation and broken eightfold symmetry from self-consistency alone [4]. Of course, with only strong interactions, the orientation of the asymmetry in the unitary space cannot be specified; one hopes that in some way the selection of specific components of the F-spin by electromagnetism and the weak interactions determines the choice of isotopic spin and hypercharge directions.

Even if we consider the scattering amplitudes of strongly interacting particles on the mass shell only and treat the matrix elements of the weak, electromagnetic, and gravitational interactions by means of dispersion theory, there are still meaningful and important questions regarding the algebraic properties of these interactions that have so far been discussed only by abstracting the properties from a formal field theory model based on fundamental entities [3] from which the baryons and mesons are built up.

If these entities were octets, we might expect the underlying symmetry group to be SU(8) instead of SU(3); it is therefore tempting to try to use unitary triplets as fundamental objects. A unitary triplet t consists of an isotopic singlet s of electric charge z (in units of e) and an isotopic doublet (u, d) with charges z+1 and z respectively. The anti-triplet $\bar{t}$ has, of course, the opposite signs of the charges. Complete symmetry among the members of the triplet gives the exact eightfold way, while a mass difference, for example, between the isotopic doublet and singlet gives the first-order violation.

For any value of z and of triplet spin, we can construct baryon octets from a basic neutral baryon singlet b by taking combinations $(b t \bar{t})$, $(b t t \bar{t} \bar{t})$, etc. **. From $(b t \bar{t})$, we get the representations **1** and **8**, while from $(b t t \bar{t} \bar{t})$ we get **1**, **8**, **10**, $\overline{\mathbf{10}}$, and **27**. In a similar way, meson singlets and octets can be made out of $(t \bar{t})$, $(t t \bar{t} \bar{t})$, etc. The quantum number $n_t - n_{\bar{t}}$ would be zero for all known baryons and mesons. The most interesting example of such a model is one in which the triplet has spin $\frac{1}{2}$ and $z = -1$, so that the four particles d^-, s^-, u^0 and b^0 exhibit a parallel with the leptons.

A simpler and more elegant scheme can be constructed if we allow non-integral values for the charges. We can dispense entirely with the basic baryon b if we assign to the triplet t the following properties: spin $\frac{1}{2}$, $z = -\frac{1}{3}$, and baryon number $\frac{1}{3}$. We then refer to the members $u^{\frac{2}{3}}$, $d^{-\frac{1}{3}}$, and $s^{-\frac{1}{3}}$ of the triplet as "quarks" [6] q and the members of the anti-triplet as anti-quarks $\bar{q}$. Baryons can now be constructed from quarks by using the combinations $(q q q)$, $(q q q q \bar{q})$, etc., while mesons are made out of $(q \bar{q})$, $(q q \bar{q} \bar{q})$, etc. It is assuming that the lowest baryon configuration $(q q q)$ gives just the representations **1**, **8**, and **10** that have been observed, while the lowest meson configuration $(q \bar{q})$ similarly gives just **1** and **8**.

A formal mathematical model based on field theory can be built up for the quarks exactly as for p, n, Λ in the old Sakata model, for example [3] with all strong interactions ascribed to a neutral vector meson field interacting symmetrically with the three particles. Within such a framework, the electromagnetic current (in units of e) is just

$$i\{\tfrac{2}{3}\,\bar{u}\,\gamma_\alpha\,u - \tfrac{1}{3}\,\bar{d}\,\gamma_\alpha\,d - \tfrac{1}{3}\,\bar{s}\,\gamma_\alpha\,s\}$$

or $\mathscr{F}_{3\alpha} + \mathscr{F}_{8\alpha}/\sqrt{3}$ in the notation of ref. [3]. For the weak current, we can take over from the Sakata model the form suggested by Gell-Mann and Lévy [7], namely $i\,\bar{p}\,\gamma_\alpha(1+\gamma_5)(n\cos\theta + \Lambda\sin\theta)$, which gives in the quark scheme the expression ***

$$i\,\bar{u}\,\gamma_\alpha(1+\gamma_5)(d\cos\theta + s\sin\theta)$$

* Work supported in part by the U.S. Atomic Energy Commission.

** This is similar to the treatment in ref. [1]. See also ref. [5].

*** The parallel with $i\,\bar{\nu}_e\,\gamma_\alpha(1+\gamma_5)\,e$ and $i\,\bar{\nu}_\mu\,\gamma_\alpha(1+\gamma_5)\mu$ is obvious. Likewise, in the model with d^-, s^-, u^0, and b^0 discussed above, we would take the weak current to be $i(\bar{b}^0\cos\theta + \bar{u}^0\sin\theta)\,\gamma_\alpha(1+\gamma_5)\,s^- + i(\bar{u}^0\cos\theta - \bar{b}^0\sin\theta)\,\gamma_\alpha(1+\gamma_5)\,d^-$. The part with $\Delta(n_t - n_{\bar{t}}) = 0$ is just $i\,\bar{u}^0\,\gamma_\alpha(1+\gamma_5)(d^-\cos\theta + s^-\sin\theta)$.

152

Volume 8, number 3 PHYSICS LETTERS 1 February 1964

or, in the notation of ref. [3],

$$[\mathscr{F}_{1\alpha} + \mathscr{F}_{1\alpha}^5 + i(\mathscr{F}_{2\alpha} + \mathscr{F}_{2\alpha}^5)] \cos\theta$$
$$+ [\mathscr{F}_{4\alpha} + \mathscr{F}_{4\alpha}^5 + i(\mathscr{F}_{5\alpha} + \mathscr{F}_{5\alpha}^5)] \sin\theta .$$

We thus obtain all the features of Cabibbo's picture [8] of the weak current, namely the rules $|\Delta I| = 1$, $\Delta Y = 0$ and $|\Delta I| = \frac{1}{2}$, $\Delta Y/\Delta Q = +1$, the conserved $\Delta Y = 0$ current with coefficient $\cos\theta$, the vector current in general as a component of the current of the F-spin, and the axial vector current transforming under SU(3) as the same component of another octet. Furthermore, we have [3] the equal-time commutation rules for the fourth components of the currents:

$$[\mathscr{F}_{j4}(x) \pm \mathscr{F}_{j4}^5(x),\ \mathscr{F}_{k4}(x') \pm \mathscr{F}_{k4}^5(x')] =$$
$$- 2 f_{jkl} [\mathscr{F}_{l4}(x) \pm \mathscr{F}_{l4}^5(x)]\, \delta(x-x') ,$$
$$[\mathscr{F}_{j4}(x) \pm \mathscr{F}_{j4}^5(x),\ \mathscr{F}_{k4}(x') \mp \mathscr{F}_{k4}^5(x')] = 0 ,$$

$i = 1, \ldots 8$, yielding the group SU(3) × SU(3). We can also look at the behaviour of the energy density $\theta_{44}(x)$ (in the gravitational interaction) under equal-time commutation with the operators $\mathscr{F}_{j4}(x') \pm \mathscr{F}_{j4}{}^5(x')$. That part which is non-invariant under the group will transform like particular representations of SU(3) × SU(3), for example like $(3, \bar{3})$ and $(\bar{3}, 3)$ if it comes just from the masses of the quarks.

All these relations can now be abstracted from the field theory model and used in a dispersion theory treatment. The scattering amplitudes for strongly interacting particles on the mass shell are assumed known; there is then a system of linear dispersion relations for the matrix elements of the weak currents (and also the electromagnetic and gravitational interactions) to lowest order in these interactions. These dispersion relations, unsubtracted and supplemented by the non-linear commutation rules abstracted from the field theory, may be powerful enough to determine all the matrix elements of the weak currents, including the effective strengths of the axial vector current matrix elements compared with those of the vector current.

It is fun to speculate about the way quarks would behave if they were physical particles of finite mass (instead of purely mathematical entities as they would be in the limit of infinite mass). Since charge and baryon number are exactly conserved, one of the quarks (presumably $u^{\frac{2}{3}}$ or $d^{-\frac{1}{3}}$) would be absolutely stable *, while the other member of the doublet would go into the first member very slowly by β-decay or K-capture. The isotopic singlet quark would presumably decay into the doublet by weak interactions, much as Λ goes into N. Ordinary matter near the earth's surface would be contaminated by stable quarks as a result of high energy cosmic ray events throughout the earth's history, but the contamination is estimated to be so small that it would never have been detected. A search for stable quarks of charge $-\frac{1}{3}$ or $+\frac{2}{3}$ and/or stable di-quarks of charge $-\frac{2}{3}$ or $+\frac{1}{3}$ or $+\frac{4}{3}$ at the highest energy accelerators would help to reassure us of the non-existence of real quarks.

These ideas were developed during a visit to Columbia University in March 1963; the author would like to thank Professor Robert Serber for stimulating them.

References

1) M.Gell-Mann, California Institute of Technology Synchrotron Laboratory Report CTSL-20 (1961).
2) Y.Ne'eman, Nuclear Phys. 26 (1961) 222.
3) M.Gell-Mann, Phys.Rev. 125 (1962) 1067.
4) E.g.: R.H.Capps, Phys.Rev. Letters 10 (1963) 312; R.E.Cutkosky, J.Kalckar and P.Tarjanne, Physics Letters 1 (1962) 93; E.Abers, F.Zachariasen and A.C. Zemach, Phys. Rev. 132 (1963) 1831; S.Glashow, Phys.Rev. 130 (1963) 2132; R.E.Cutkosky and P.Tarjanne, Phys.Rev. 132 (1963) 1354.
5) P.Tarjanne and V.L.Teplitz, Phys.Rev. Letters 11 (1963) 447.
6) James Joyce, Finnegan's Wake (Viking Press, New York, 1939) p.383.
7) M.Gell-Mann and M.Lévy, Nuovo Cimento 16 (1960) 705.
8) N.Cabibbo, Phys.Rev. Letters 10 (1963) 531.

* There is the alternative possibility that the quarks are unstable under decay into baryon plus anti-di-quark or anti-baryon plus quadri-quark. In any case, some particle of fractional charge would have to be absolutely stable.

* * * * *

Light Cone Current Algebra

Harald Fritzsch

H. Fritzsch (✉)
Physik-Department, Ludwig-Maximilians-Universität Physik-Department, München, Germany
e-mail: fritzsch@mppmu.mpg.de

H. Fritzsch (ed.), *Murray Gell-Mann and the Physics of Quarks*, Classic Texts in the Sciences, https://doi.org/10.1007/978-3-319-92195-2_6

9

Light Cone Current Algebra*

HARALD FRITZSCH** and MURRAY GELL-MANN†

PREFACE

THIS TALK follows by a few months a talk by the same authors on nearly the same subject at the Coral Gables Conference. The ideas presented here are basically the same, but with some amplification, some change of viewpoint, and a number of new questions for the future. For our own convenience, we have transcribed the Coral Gables paper, but with an added ninth section, entitled "Problems of light cone current algebra", dealing with our present views and emphasizing research topics that require study.

1. INTRODUCTION

We should like to show that a number of different ideas of the last few years on broken scale invariance, scaling in deep inelastic electron-nucleon scattering, operator product expansions on the light cone, "parton" models, and generalizations of current algebra, as well as some new ideas, form a coherent picture. One can fit together the parts of each approach that make sense and obtain a consistent view of scale invariance, broken by certain terms in the energy density, but restored in operator commutators on the light cone.

We begin in the next section with a review of the properties of the dilation operator D obtained from the stress-energy-momentum tensor $\theta_{\mu\nu}$ and the behavior of operators under equal-time

* Work supported in part by the U.S. Atomic Energy Commission under contract AT(11-1)-68, San Francisco Operations Office.

** Max-Planck-Institut für Physik und Astrophysik, München, Germany. Present address (1971-1972): CERN, Geneva, Switzerland.
† California Institute of Technology, Pasadena, California. Present address: (1971-1972): CERN, Geneva, Switzerland.

commutation with D, which is described in terms of physical dimensions l for the operators. We review the evidence on the relation between the violation of scale invariance and the violation of $SU \times SU_3$ invariance.

Next, in Section 3, we describe something that may seem at first sight quite different, namely the Bjorken scaling of deep inelastic scattering cross sections of electrons on nucleons and the interpretation of this scaling in terms of the light cone commutator of two electromagnetic current operators. We use a generalization of Wilson's work,[1] the light-cone expansion emphasized particularly by Brandt and Preparata[2] and Frishman.[3] A different definition l of physical dimension is thus introduced and the scaling implies a kind of conservation of l on the light cone. On the right-hand side of the expansions, the operators have $l = -J - 2$, where J is the leading angular momentum contained in each operator and l is the leading dimension.

In Section 4, we show that under simple assumptions the dimensions l and l are essentially the same, and that the notions of scaling and conservation of dimension can be widely generalized. The essential assumption of the whole approach is seen to be that the dimension l (or l) of any symmetry-breaking term in the energy (whether violating scale invariance or $SU_3 \times SU_3$) is *higher* than the dimension, -4, of the completely invariant part of the energy density. The conservation of dimension on the light cone then assigns a lower singularity to symmetry-breaking terms than to symmetry-preserving terms, permitting the light-cone relations to be completely symmetrical under scale, $SU_3 \times SU_3$, and perhaps other symmetries.

In Section 5, the power series expansion on the light cone is formally summed to give bilocal operators (as briefly discussed by Frishman) and it is suggested that these bilocal light-cone operators may be very few in number and may form some very simple closed algebraic system. They are then the basic mathematical entities of the scheme.

It is pointed out that several features of the Stanford experiments, as interpreted according to the ideas of scaling, resemble the behavior on the light cone of free field theory or of interacting field theory with naïve manipulation of operators, rather than the behavior of renormalized perturbation expansions of renormalizable field theories. Thus free field theory models may be studied for the purpose of abstracting algebraic relations that might be true on the light cone in the real world of hadrons. (Of course, matrix elements of operators in the real world would not in general resemble matrix elements in free field theory.) Thus in Section 6 we study the light-cone behavior of local and bilocal operators in free quark theory, the simplest interesting case. The relevant bilocal operators turn out to be extremely simple, namely just $i/2(\bar{q}(x)\lambda_i\gamma_\alpha q(y))$ and $i/2(\bar{q}(x)\lambda_i\gamma_\alpha\gamma_5 q(y))$, bilocal generalizations of V and A currents. The algebraic system to which they belong is also very simple.

In Section 7 we explore briefly what it would mean if these algebraic relations of free quark theory were really true on the light cone for hadrons. We see that we obtain, among other things, the sensible features of the so-called "parton" picture of Feynman[4] and of Bjorken and Paschos,[5] especially as formulated more exactly by Landshoff and Polkinghorne,[6] Llewellyn Smith,[7] and others. Many symmetry relations are true in such a theory, and can be checked by deep inelastic experiments with electrons and with neutrinos. Of course, some alleged results of the "parton" model depend not just on light cone commutators but on detailed additional assumptions about matrix elements, and about such results we have nothing to say.

The abstraction of free quark light cone commutation relations becomes more credible if we can show, as was done for equal time charge density commutation relations, that certain kinds of non-trivial interactions of quarks leave the relations undisturbed, according to the method of naïve manipulation of operators, using equations of motion. There is evidence that in fact this is so, in a theory with a neutral scalar or pseudoscalar "gluon" having a Yukawa interaction with the quarks. (If the "gluon" is a vector boson, the commutation relations on the light cone might be disturbed for all we know.)

A special case is one in which we abstract from a model in which there are only quarks, with some unspecified self-interaction, and no "gluons". This corresponds to the pure quark case of the "parton" model. One additional constraint is added, namely the identification of the traceless part of $\theta_{\mu\nu}$ with the analog of the traceless part of the symmetrized $\bar{q}\gamma_\mu\partial_\nu q$. This constraint leads to an additional sum rule for deep inelastic electron and neutrino experiments, a rule that provides a real test of the pure quark case.

We do not, in this paper, study the connection between scaling in electromagnetic and neutrino experiments on hadrons on the one hand and scaling in "inclusive" reactions of hadrons alone on the other hand. Some approaches, such as the intuition of the "parton" theorists, suggest such a connection, but we do not explore that idea here. It is worth reemphasizing, however, that any theory of pure hadron behavior that limits transverse momenta of particles produced at high energies has a chance of giving the Bjorken scaling when electromagnetism and weak interactions are introduced. (This point has been made in the cut-off models of Drell, Levy, and Yan.[8])

2. DILATION OPERATOR AND BROKEN SCALE INVARIANCE[9]

We assume that gravity theory (in first order perturbation approximation) applies to hadrons on a microscopic scale, although no way of checking that assertion is known. There is then a symmetrical, conserved, local stress-energy-momentum tensor $\theta_{\mu\nu}(x)$ and in terms of it the translation operators P_μ, obeying for any operator $\mathcal{O}\cdots(x)$, the relation

$$[\mathcal{O}\cdots(x), P_\mu] = \frac{1}{i}\,\partial_\mu \mathcal{O}\cdots(x), \tag{2.1}$$

are given by

$$P_\mu = \int \theta_{\mu o} d^3x. \tag{2.2}$$

Now we want to define a differential dilation operator $D(t)$ that corresponds to our intuitive notions of such an operator, i.e., one that on equal-time commutation with a local operator $\mathcal{O}\cdots$ of definite physical dimension l_o, gives

$$[\mathcal{O}\cdots(x), D(t)] = ix_\mu\partial_\mu\mathcal{O}\cdots(x) - il_\sigma\mathcal{O}\cdots(x). \tag{2.3}$$

We suppose that gravity selects a $\theta_{\mu\nu}$ such that this dilation operator D is given by the expression

$$D = -\int x_\mu \theta_{\mu o} d^3x. \tag{2.4}$$

It is known that for any renormalizable theory this is possible, and Callan, Coleman, and Jackiw have shown that in such a case the matrix elements of this $\theta_{\mu\nu}$ are finite. From (2.4) we see that the violation of scale invariance is connected with the non-vanishing of $\theta_{\mu\mu}$, since we have

$$\frac{dD}{dt} = -\int \theta_{\mu\mu} d^3x. \tag{2.5}$$

Another version of the same formula says that

$$[D, P_0] = -iP_0 - i\int \theta_{\mu\mu} d^3x \tag{2.6}$$

and we see from this and (2.3) that the energy denisty has a main scale-invariant term $\bar{\bar{\theta}}_{00}$ (under the complete dilation operator D) with $l = -4$ (corresponding to the mathematical dimension of energy density) and other terms w_n with other physical dimensions l_n. The simplest assumption (true of most simple models) is that these terms are world scalars, in which case we obtain

$$-\theta_{\mu\mu} = \sum_n (l_n + 4) w_n \tag{2.7}$$

along with the definition

$$\theta_{00} = \bar{\bar{\theta}}_{00} + \sum_n w_n. \tag{2.8}$$

We note that the breaking of scale invariance prevents D from being a world scalar and that equal-time commutation with D leads to a non-covariant break-up of operators into pieces with different dimensions l.

To investigate the relation between the violations of scale invariance and of chiral invariance, we make a still further simplifying assumption (true of many simple models such as the quark-gluon Langrangian model), namely that there are two q-number w's, the first violating scale invariance but not chiral invariance (like the gluon mass) and the second violating both (like the quark mass):

$$\theta_{00} = \bar{\bar{\theta}}_{00} + \delta + u + \text{const.}, \tag{2.9}$$

with δ transforming like $(\mathbf{1}, \mathbf{1})$ under $SU_3 \times SU_3$. Now how does

u transform? We shall start with the usual theory that it all belongs to a single $(3, \bar{3}) + (\bar{3}, 3)$ representation and that the smallness of m_π^2 is to be attributed, in the spirit of PCAC, to the small violation of $SU_2 \times SU_2$ invariance by u. In that case we have

$$u = -u_0 - cu_8, \tag{2.10}$$

with c not far from $-\sqrt{2}$, the value that gives $SU_2 \times SU_2$ invariance and $m_\pi^2 = 0$ and corresponds in a quark scheme to giving a mass only to the s quark. A small amount of u_3 may be present also, if there is a violation of isotopic spin conservation that is not directly electromagnetic; an expression containing u_0, u_3, and u_8 is the most general canonical form of a CP-conserving term violating $SU_3 \times SU_3$ invariance and transforming like $(3, \bar{3}) + (\bar{3}, 3)$.

According to all these simple assumptions, we have

$$-\theta_{\mu\mu} = (l_\delta + 4)\delta + (l_u + 4)(-u_0 - cu_8) + 4\,(\text{const.}) \tag{2.11}$$

and, since the expected value of $(-\theta_{\mu\mu})$ is $2m^2$, we have

$$O = (l_\delta + 4)\langle \text{vac}|\delta|\text{vac}\rangle + (l_u + 4)\langle \text{vac}|u|\text{vac}\rangle + 4\,(\text{const.}), \tag{2.12}$$

$$2m_i^2(PS\,8) = (l_\delta + 4)(PS_i|\delta|PS_i) + (l_u + 4)\langle PS_i|u|PS_i\rangle, \tag{2.13}$$

etc.

The question has often been raised whether δ could vanish. Such a theory is very interesting, in that the same term u would break chiral and conformal symmetry. But is it possible?

It was pointed out a year or two ago[10] that for this idea to work, something would have to be wrong with the final result of von Hippel and Kim,[11] who calculated approximately the "σ terms" in meson-baryon scattering and found, using our theory of $SU_3 \times SU_3$ violation, that $\langle N|u|N\rangle$ was very small compared to $2m_N^2$. Given the variation of $\langle B|u_8|B\rangle$ over the $1/2^+$ baryon octet, the ratio of $\langle \Xi|u|\Xi\rangle$ to $\langle N|u|N\rangle$ would be huge if von Hippel and Kim were right, and this disagrees with the value m_Ξ^2/m_N^2 that obtains if $\delta = 0$.

Now, Ellis[12] has shown that in fact the method of von Hippel and Kim should be modified and will produce different results, provided there is a dilation. A dilation is a neutral scalar meson that dominates the dispersion relations for matrix elements of $\theta_{\mu\mu}$ at low frequency, just as the pseudoscalar octet is supposed to dominate the relations for $\partial_\alpha \mathscr{F}^5_{i\alpha}$. We are dealing in the case of the dilaton, with PCDC (partially conserved dilation current) along with PCAC (partially conserved axial vector current). If we have PCAC, PCDC, *and* $\delta = 0$, we may crudely describe the situation by saying that as $u \to 0$ we have chiral and scale invariance of the energy, the masses of a pseudoscalar octet and a scalar singlet go to zero, and the vacuum is not invariant under either chiral or scale transformations (though it is probably SU_3 invariant). With the dilation, we can have masses of other particles non-vanishing as $u \to 0$, even though that limit is scale invariant.

Dashen and Cheng[13] have just finished a different calculation of the "σ terms" not subject to modification by dilation effects, and they find, using our description of the violation of chiral invariance, that $\langle N|u|N\rangle$ at rest is around $2m_N^2$, a result perfectly compatible with the idea of vanishing δ and yielding in that case a value $l_u \approx -3$ (as in a naive quark picture, where u is a quark mass term!).

An argument was given last year [10] that if $\delta = 0$, the value of l_u would have to be -2 in order to preserve the perturbation theory approach for $m^2(PS\,8)$, $m^2(PS\,8) \propto u$, which gives the right mass formula for the pseudoscalar octet. Ellis, Weisz, and Zumino[14] have shown that this argument can be evaded if there is a dilation.

Thus at present there is nothing known against the idea that $\delta = 0$, with l_u probably equal to -3. However, there is no strong evidence in favor of the idea either. Theories with non-vanishing δ operators and various values of l_δ and l_u are not excluded at all (although even here a dilaton would be useful to explain why $\langle N|u|N\rangle$ is so large). It is a challenge to theorists to propose experimental means of checking whether the δ operator is there or not.

It is also possible that the simple theory of chiral symmetry violation may be wrong. First of all, the expression $-u_0 + \sqrt{2}u_8$

could be right for the $SU_2 \times SU_2$-conserving but $SU_3 \times SU_3$-violating part of θ_{00}, while the $SU_2 \times SU_2$ violation could be accomplished by something quite different from $(-c-\sqrt{2})u_8$. Secondly, there can easily be an admixture of the eighth component g_8 of an octet belonging to $(1, 8)$ and $(8, 1)$. Thirdly, the whole idea of explaining $m_\pi^2 \approx 0$ by near-conservation of $SU_2 \times SU_2$ might fail, as might the idea of octet violation of SU_3; it is those two hypotheses that give the result that for $m_\pi^2 = 0$ we have only $u_o - \sqrt{2}u_8$ with a possible admixture of g_8. Here again there is a challenge to theoreticians to propose effective experimental tests of the theory of chiral symmetry violation.

3. LIGHT CONE COMMUTATORS AND DEEP INELASTIC ELECTRON SCATTERING

We want ultimately to connect the above discussion of physical dimensions and broken scale invariance with the scaling described in connection with the Stanford experiments on deep inelastic electron scattering.[15] We must begin by presenting the Stanford scaling in suitable form. For the purpose of doing so, we shall assume for convenience that the experiments support certain popular conclusions, even though uncertainties really prevent us from saying more than that the experiments are consistent with such conclusions:

(1) that the scaling formula of Bjorken is really correct, with no logarithmic factors, as the energy and virtual photon mass go to infinity with fixed ratio;

(2) that in this limit the neutron and proton behave differently;

(3) that in the limit the longitudinal cross section for virtual photons goes to zero compared to the transverse cross section.

All these conclusions are easy to accept if we draw our intuition from certain field theories without interactions or from certain field theories with naïve manipulation of operators. However, detailed calculations using the renormalized perturbation expansion in renormalizable field theories do not reveal any of these forms of behavior, unless of course the sum of all orders of perturbation theory somehow restores the simple situation. If we accept the conclusions, therefore, we should probably not think in terms of the renormalized perturbation expansion, but rather conclude, so to speak, that Nature reads books on free field theory, as far as the Bjorken limit is concerned.

To discuss the Stanford results, we employ a more or less conventional notation. The structure functions of the nucleon are defined by matrix elements averaged over nucleon spin,

$$\begin{aligned}\frac{1}{4\pi}\int d^4x\langle N,p|[j_\mu(x),j_\nu(y)]|N,p\rangle e^{-iq\cdot(x-y)}\\ &= \left(\delta_{\mu\nu}-\frac{q_\mu q_\nu}{q^2}\right)W_1(q^2,p\cdot q)\\ &+\left(p_\mu-\frac{p\cdot q}{q^2}q_\mu\right)\left(p_\nu-\frac{p\cdot q}{q^2}q_\nu\right)W_2(q^2,p\cdot q)\\ &= \left(\delta_{\mu\nu}-\frac{q_\mu q_\nu}{q^2}\right)\left(W_1-\frac{(p\cdot q)^2}{q^2}W_2\right.\\ &+\frac{\delta_{\mu\nu}(p\cdot q)^2+p_\mu p_\nu q^2-(p_{\mu\nu}+q_\mu p_\nu)p\cdot q}{q^2}W_2,\end{aligned} \tag{3.1}$$

where p is the nucleon four-momentum and q the four-momentum of the virtual photon. As q^2 and $q\cdot p$ become infinite with fixed ratio, averaging over the nucleon spin and assuming $\sigma_L/\sigma_T\mathscr{F}\to 0$, we can write the Bjorken scaling in the form

$$\begin{aligned}\frac{1}{4\pi}\int d^4x\langle N,p|[j_\mu(x),j_\nu(y)]|N,p\rangle e^{-iq\cdot(x-y)}\\ \to\frac{(p_\mu q_\nu+p_\nu q_\mu)p\cdot q-\delta_{\mu\nu}(p\cdot q)^2-p_\mu p_\nu q^2}{q^2(q\cdot p)}F_2(\xi),\end{aligned} \tag{3.2}$$

where $\xi=-q^2/2p\cdot q$ and $F_2(\xi)$ is the scaling function in the deep inelastic region.

In coordinate space, this limit is achieved by approaching the light cone $(x-y)^2=0$,[15] and we employ a method, used by Frishman[3] and by Brandt and Preparata,[2] generalizing earlier work of Wilson, that starts with an expansion for commutators or operator products valid near $(x-y)^2=0$. (The symbol $\hat{=}$ will be employed for equality in the vicinity of the light cone.) After the expansion is made, then the matrix element is taken between nucleons. To simplify matters, let us introduce the "barred product" of two operators, which means that we average over the mean position $R \equiv (x + y)/2$, leaving a function of $z \equiv x - y$ only (as appropriate for matrix elements with no change of momentum) and that we retain in the expansion only totally symmetric Lorentz tensor operators (as appropriate for matrix elements averaged over spin). Then the assumed light-cone expansion of the barred commutator $[\overline{j_\mu(x), j_\nu(y)}]$ tells us that we have, as $z^2 \to 0$,

$$[\overline{j_\mu(x), j_\nu(y)}] \hat{=} t_{\mu\nu\rho\sigma}\{\varepsilon(z_0)\delta(z^2)(\mathcal{O}_{\rho\sigma} + \frac{1}{2!} z_\alpha z_\beta \mathcal{O}_{\rho\sigma\alpha\beta} + \cdots)\}$$
$$+ (\partial_\mu\partial_\nu - \delta_{\mu\nu}\partial^2)\{\varepsilon(z_0)\delta(z^2)(U + \frac{1}{2!} z_\alpha z_\beta U_{\alpha\beta} + \cdots)\}, \tag{3.3}$$

where

$$t_{\mu\nu\rho\sigma} = \frac{1}{\pi i} \times$$
$$\frac{2\delta_{\mu\nu}\partial_\rho\partial_\sigma - \delta_{\rho\mu}\partial_\nu\partial_\sigma - \delta_{\rho\nu}\partial_\mu\partial_\sigma - \delta_{\sigma\mu}\partial_\nu\partial_\rho - \delta_{\sigma\nu}\partial_\mu\partial_\sigma - \delta_{\mu\sigma}\delta_{\nu\rho}\partial^2 - \partial_{\nu\sigma}\delta_{\mu\rho}\partial^2}{\partial^2}$$

and the second term, the one that gives σ_L, will be ignored for simplicity in our further work.

In order to obtain the Bjorken limit, we have only to examine the matrix elements between $|Np\rangle$ and itself of the operators $\mathcal{O}_{\alpha\beta}$, $\mathcal{O}_{\alpha\beta\gamma\delta}$, $\mathcal{O}_{\alpha\beta\gamma\delta\varepsilon\rho}$, etc. The leading tensors in the matrix elements have the form $c_2 p_\alpha p_\beta$, $c_4 p_\alpha p_\beta p_\gamma p_\delta$, etc., where the c's are dimensionless constants. The lower tensors, such as $\delta_{\alpha\beta}$, have coefficients that are positive powers of masses, and these tensors give negligible contributions in the Bjorken limit. All we need is the very weak assumption that c_2, c_4, c_8, etc., are not all zero, and we obtain the Bjorken limit.

We define the function

$$\tilde{F}(p \cdot z) = c_2 + \frac{1}{2!} \cdot c_4(p \cdot z)^2 + \cdots. \tag{3.4}$$

Taking the Fourier transform of the matrix element of (3.3), we get in the Bjorken limit

$$\begin{aligned} W_2 &\to \frac{1}{2\pi^2 i}\int d^4z e^{-iq\cdot z}\tilde{F}(p\cdot z)\varepsilon(z_0)\delta(z^2) \\ &= \frac{1}{2\pi^2 i}\int F(\xi)d\xi \int d^4z e^{-i(q+\xi p)\cdot z}\varepsilon(z_0)\delta(z^2) \\ &= 2\int F(\xi)d\xi\varepsilon(-q\cdot p)\delta(q^2+2q\cdot p\xi) \\ &= \frac{1}{-q\cdot p}F(\xi) \end{aligned} \tag{3.5}$$

where ξ is $-q^2/2q\cdot p$ and $F(\xi)$ is the Fourier transform of $\tilde{F}(p\cdot z)$:

$$F(\xi) = \frac{1}{2\pi}\int e^{i\xi(p\cdot z)}\tilde{F}(p\cdot z)d(p\cdot z). \tag{3.6}$$

The function $F(\xi)$ is therefore the Bjorken scaling function in the deep inelastic limit and is defined only for $-1 < \xi < 1$. We can write (3.6) in the from

$$F(\xi) = c_2\cdot\delta(\xi) - c_4\frac{1}{2!}\delta''(\xi) + c_6\frac{1}{4!}\delta''''(\xi) - \cdots \tag{3.7}$$

The dimensionless numbers c_i defined by the matrix elements of the expansion operators can be written as

$$c_2 = \int_{-1}^{1} F(\xi)d\xi, \quad c_4 = -\int_{-1}^{1} F(\xi)\xi^2 d\xi\cdots \tag{3.8}$$

This shows the connection between the matrix elements of the expansion operators and the moments of the scaling function. The Bjorken limit is seen to be a special case (the matrix element between single nucleon states of fixed momentum) of the light cone expansion.[17]

Now the derivation of the Bjorken limit from the light cone expansion can be described in terms of a kind of physical dimension l for operators. (We shall see in the next section that these dimensions l are essentially the same as the physical dimensions l we described in Section 2.) We define the expansion to conserve dimension on the light cone and assign to each current

$l = -3$ while counting each power of z as having an l-value equal to the power. We see then that on the right-hand side we are assigning to each J-th rank Lorentz tensor (with maximum spin J) the dimension $l = -J-2$. Furthermore, the physical dimension equals the mathematical dimension in all of these cases.

4. GENERALIZED LIGHT CONE SCALING AND BROKEN SCALE INVARIANCE

We have outlined a situation in which scale invariance is broken by a non-vanishing $\theta_{\mu\mu}$ but restored in the most singular terms of current commutators on the light cone. There is no reason to suppose that such a restoration is restricted to commutators of electromagnetic currents. We may extend the idea to all the vector currents $\mathscr{F}_{i\mu}$ and axial vector currents $\mathscr{F}_{i\mu}{}^5$, to the scalar and pseudoscalar operators u_i and v_i that comprise the $(3, \bar{3})$ and $(\bar{3}, 3)$ representation thought to be involved in chiral symmetry breaking, to the whole stress-energy momentum tensor $\theta_{\mu\nu}$, to any other local operator of physical significance, and finally to all the local operators occurring in the light cone expansions of commutators of all these quantities with one another. Let us suppose that in fact conservation of dimension applies to leading terms in the light cone in the commutators of all these quantities and that finally a closed algebraic system with an infinite number of local operators is attained, such that the light cone commutator of any two of the operators is expressible as a linear combination of operators in the algebra. We devote this section and the next one to discussing such a situation.

If there is to be an analog of Bjorken scaling in all these situations, then on the right-hand side of the light cone commutation relations we want operators with $l = -J-2$, as above for electromagnetic current commutators, so that we get leading matrix elements between one-particle states going like $cp_\alpha p_\beta \cdots$, where the c's are dimensionless constants.

Of course, there might be cases in which, for some reason, all the c's have to vanish, and the next-to-leading term on the light cone becomes the leading term. Then the coefficients would have the dimensions of positive powers of mass. We want to avoid, however, situations in which coefficients with the dimension of negative powers of mass occur; that means on the right-hand side we want $l \leqq -J-2$ in any case, and $l = -J-2$ when there is nothing to prevent it.

This idea might have to be modified, as in a quark model with a scalar or pseudoscalar "gluon" field, to allow for a single operator ϕ, with $l = -1$ and $J = 0$, that can occur in a barred product, but without a sequence of higher tensors with $l = -J-1$ that could occur in such a product; gradients of ϕ would, of course, average out in a barred product. However, even this modification is probably unnecessary, since preliminary indications are that, in the light cone commutator of any two (physically interesting operators, the operator ϕ with $l = -1$ would not appear on the right-hand side.

Now, on the left-hand side, we want the non-conserved currents among $\mathscr{F}_{i\mu}$ and $\mathscr{F}_{i\mu}^5$ to act as if they have dimension -3 just like the conserved ones, as far as leading singularities on the light cone are concerned, even though the non-conservation implies the admixture of terms that may have other dimensions l, dimensions that become $l-1$ in the divergences, and correspond to dimensions $l-1$ in the $SU_3 \times SU_3$ breaking terms in the energy density. But the idea of conservation of dimension on the light cone tells us that we are dealing with lower singularities when the dimensions of the operators on the left are greater. What is needed, then, is for the dimensions l to be > -3, i.e., for the chiral symmetry breaking terms in $\theta_{\mu\nu}$ to have dimension > -4. Likewise, if we want the stress-energy-momentum tensor itself to obey simple light cone scaling, we need to have the dimension of all scale breaking parts of $\theta_{\mu\nu}$ restricted to values > -4. In general, we can have symmetry on the light cone if the symmetry breaking terms in $\theta_{\mu\nu}$ have dimension greater than -4. (See Appendix I.)

Now we can have $\mathscr{F}_{i\mu}$ and $\mathscr{F}_{i\mu}^5$ behaving, as far as leading singularities on the light cone are concerned, like conserved currents with $l = -3$, $\theta_{\mu\nu}$ behaving like a chiral and scale invariant quantity with $l = -4$, and so forth. To pick out the subsidiary dimensions associated with the non-conservation of $SU_3 \times SU_3$ and dilation, we can study light cone commutators involving,

$\partial_\alpha \mathscr{F}_{l\alpha}$, $\partial_\alpha \mathscr{F}^5_{i\alpha}$, and $\theta_{\mu\mu}$. (If the $(3, \bar{3}) + (\bar{3}, 3)$ hypothesis is correct, that means studying commutators involving u's and v's and also δ, if $\delta \neq 0$.)

In our enormous closed light cone algebra, we have all the operators under consideration occuring on the left-hand side, the ones with $l = -J - 2$ on the right-hand side, and coefficients that are functions of z behaving like powers according to the conservation of dimension. But are there restrictions on these powers? And are there restrictions on the dimensions occurring among the operators?

If, for example, the functions of z have to be like powers of z^2 (or $\delta(z^2)$, $\delta'(z^2)$, etc.) multiplied by tensors $z_\alpha z_\beta z_\gamma \cdots$, and if $l + J$ for some operators is allowed to be non-integral or even odd integral, then we cannot always have $l = -J - 2$ on the right, i.e., the coefficients of all such operators would vanish in certain commutators, and for those commutators we would have to be content with operators with $l < -J - 2$ on the right, and coefficients of leading tensors that act like positive powers of a mass.

Let us consider the example:

$$[\theta_{\mu\nu}(x), u(y)] \mathrel{\hat{=}} E_{\mu\nu}(z) \cdot (\mathcal{O}(y) + z_\rho \mathcal{O}_\rho(y) + \cdots) + \cdots,$$

where $u(y)$ has the dimension -3. In this case we cannot have the Bjorken scaling. Because of the relation

$$[D(O), u(O)] = -3iu(O),$$

the operator $\mathcal{O}(y)$ has to be proportional to $u(y)$. The operator series fulfilling the condition $l = -J - 2$ is forbidden in this case on the right-hand side.

We have already emphasized that Nature seems to imitate the algebraic properties of free field theory rather than renormalized perturbation theory. (We could also say that Nature is imitating a super-renormalizable theory, even though no sensible theory of that kind exists, with the usual methods of renormalization, in four dimensions.) This suggests that we should have in our general expansion framework finite equal-time commutators for all possible operators and their time derivatives.

Such a requirement means that all functions of z multiplying operators in a light cone expansion must have the behavior described just above, i.e., the scalar functions involved behave like integral forces of z^2 or like derivatives of delta functions with z^2 as the argument. The formula

$$\frac{1}{(z^2+i\varepsilon)^\alpha} - \frac{1}{(z^2-i\varepsilon)^\alpha} \xrightarrow[z_0\to 0]{} \text{const. } z_0^{-2\alpha+3}\,\delta(z)$$

shows the sort of thing we mean. It also shows that α must not be too large. That can result in lower limits on the tensorial rank of the first operator in the light cone expansion in higher and higher tensors; to put it differently, the first few operators in a particular light cone expansion may have to be zero in order to give finiteness of equal time commutators with all time derivatives.

Now, on the right-hand side of a light cone commutator of two physically interesting operators, when rules such as we have just discussed do not forbid it, we obtain operators with definite $SU_3 \times SU_3$ and other symmetry properies, of various tensor ranks, and with $l = -J-2$. Now, for a given set of quantum numbers, how many such operators are there? Wilson[1] suggested a long time ago that there may be very few, sometimes only one, and others none. Thus no matter what we have on the left, we always would get the same old operators on the right (when not forbidden and less singular terms with dimensional coefficients occuring instead). This is very important, since the matrix elements of these universal $l = -J-2$ operators are then natural constants occurring in many problems. Wilson presumably went a little too far in guessing that the only Lorentz tensor operator in the light cone expansion of $[\overline{j_\mu(x),\ j_\nu(y)}]$ would be the stress-energy-momentum tensor $\theta_{\mu\nu}$, with no provision for an accompanying octet of $l = -4$ tensors. That radical suggestion, as shown by Mack,[17] would make $\int F_2^{en}(\xi)\,d\xi$ equal to $\int F_2^{ep}(\xi) d\xi$, which does not appear to be the case. However, it is still possible that one singlet and one octet of tensors may do the job. (See the discussion in Section 7 of the "pure quark" case.)

If we allow z_0 to approach zero in a light cone commutator, we obtain an equal time commutator. If Wilson's principle (suitably weakened) is admitted, then all physically interesting operators must obey some equal time commutation relations, with well-known operators on the right-hand side, and presumably there are fairly small algebraic systems to which these equal time commutators belong. The dimensions of the operators constrain severely the nature of the algebra involved. For example, suppose $SU_3 \times SU_3$ is broken by a quantity u belonging to the representation $(3, \bar{3}) \oplus (\bar{3}, 3)$ and having a single dimension l_u. Then, if $l_u = -3$, we may well have the algebraic system proposed years ago by one of us (M.G.-M.) in which F_i, F_i^5, $\int u_i d^3x$, and $\int v_i d^3x$ obey the E.T.C. relations of U_6, as in the quark model. If $l_u = -2$, however, then we would have $\int u_i d^3x$ and $d/dt \int u_i d^3x$ commuting to give a set of quantities including $\int u_i d^3x$, and so forth.

We have described scaling in this section as if the dimensions l were closely related to the dimensions $\bar{l}$ obtained by equal time commutation with the dilation operator D in Section 2. Let us now demonstrate that this is so.

To take a simple case, suppose that in the light cone commutator of an operator $\mathcal{O}\cdots$ with itself, the same operator $\mathcal{O}\cdots$ occurs in the expansion on the right-hand side. Then we have a situation crudely described by the equation

$$[\mathcal{O}\cdots(z), \mathcal{O}\cdots(0)] \mathrel{\hat{=}} \cdots + (z)^l\mathcal{O}\cdots(0) + \cdots, \tag{4.1}$$

where l is the principal dimension of $\mathcal{O}\cdots$. Here $(z)^l$ means any function of z with dimension l, and we must have that because of conservation of dimension. Now under equal time commutation with D, say $\mathcal{O}\cdots$ exhibits dimension $\bar{l}$. Let $z_0 \to 0$ and perform the equal time commutation, according to Eq. (2.3). We obtain

$$\begin{aligned}(iz \cdot \nabla - 2i\bar{l})[\mathcal{O}\cdots(z), \mathcal{O}\cdots(0)] &= -i\bar{l}(z)^l\mathcal{O}\cdots(0) \\ &= (il - 2i\bar{l})(z)^l\mathcal{O}\cdots(0)\end{aligned} \tag{4.2}$$

so that $l = \bar{l}$, as we would like.

Now to generalize the demonstration, we consider the infinite closed algebra of light cone commutators, construct commutators like (4.1) involving different operators, and from commutation with D as in (4.2) obtain equations

$$l_1 + l_2 - l_3 = \bar{l}_1 + \bar{l}_2 - \bar{l}_3, \tag{4.3}$$

where $\mathcal{O}\cdots^{(1)}$ and $\mathcal{O}\cdots^{(2)}$ are commuted and yield a term containing $\mathcal{O}\cdots^{(3)}$ on the right. Chains of such relations can then be used to demonstrate finally that $l = \bar{l}$ for the various operators in which we are interested.

The subsidiary dimensions associated with symmetry breaking have not been treated here. They can be dealt with in part by isolating the expressions $\partial_\mu \mathscr{F}^5_{i\mu}$, $\theta_{\mu\mu}$, etc., that exhibit only the subsidiary dimensions and applying similar arguments to them. In that way we learn that also for subsidiary dimensions $\bar{l} = l$.

However, the subsidiary dimensions, while numerically equal for the two definitions of dimension, do not enter in the same way for the two definitions. The physical dimension l defined by light cone commutation always enters covariantly, while $\bar{l}$ is defined by equal time commutation with the quantity D and enters non-covariantly, as in the break-up of $\theta_{\mu\nu}$ into the leading term $\bar{\bar{\theta}}_{\mu\nu}$ of dimension -4 and the subsidiary ones of higher dimensions. If these others come from world scalars w_n of dimensions l_n, then we have

$$\theta_{\mu\nu} = \bar{\bar{\theta}}_{\mu\nu} + \sum_n \{(3+l)\delta_{\mu\nu} + (4+l)\delta_{\mu 0}\delta_{\nu 0}\}\frac{w_n}{3}, \tag{4.4}$$

so that we agree with the relations

$$\theta_{00} = \bar{\bar{\theta}}_{00} + \sum_n w_n, \tag{2.8}$$

$$-\theta_{\mu\mu} = \sum_n (l_n + 4) w_n. \tag{2.9}$$

Clearly, $\bar{\bar{\theta}}_{\mu\nu}$ is non-covariant.

To obtain the non-covariant formula from the covariant one, the best method is to write the light cone commutator of an operator with $\theta_{\mu\nu}$, involving physical dimensions l, and then construct $D = -\int x_\mu \theta_{\mu 0} d^3x$ out of $\theta_{\mu\nu}$ and allow the light cone commutator to approach an equal time commutator. The non-covariant formula involving $\bar{l}$ must then result.

As an example of non-covariant behavior of equal time commutation with D, consider such a commutator involving an arbitrary tensor operator $\mathcal{O}_{\rho\sigma}$ of dimension -4. We may pick up non-covariant contributions that arise from lower order terms near the light cone than those that give the dominant scaling behavior. We may have

$$[\theta_{\mu\nu}(x), \mathcal{O}_{\rho\sigma}(y)] = \text{leading term} + \partial_\mu\partial_\nu\partial_\rho\partial_\sigma\{\varepsilon(z_0)\delta(z^2)[\mathcal{O}(y) + \cdots]\} + \cdots$$

giving the result

$$\text{E.T.C. } [D, \mathcal{O}_{\rho\sigma}(0)] = 4i\theta_{\rho\sigma}(0) + \text{const.}\, \delta_{\rho 0}\delta_{\sigma 0}\mathcal{O}(0) + \cdots.$$

For commutation of D with a scalar operator, there is no analog of this situation.

5. BILOCAL OPERATORS

So far, in commuting two currents at points separated by a four-dimensional vector z_μ, we have expanded the right-hand side on the light cone in powers of z_μ. It is very convenient for many purposes to sum the series and obtain a single operator of low Lorentz tensor rank that is a function of z. In a barred commutator, it is a function of z only, but in an ordinary unbarred commutator, it is a function of z and $R \equiv (x + y)/2$, in other words, a function of x and y. We call such an operator a bilocal operator and write it as $\mathcal{O}\cdots(x, y)$ or, in barred form, $\bar{\mathcal{O}}\cdots(x, y)$.

We can, for example, write Eq. (3.3) in the form

$$\overline{[j_\mu(x), j_\nu(y)]} \mathrel{\hat{=}} t_{\mu\nu\rho\sigma}\{\varepsilon(z_0)\delta(z^2)\overline{\mathcal{O}_{\rho\sigma}(x, y)}\} + \text{longitudinal term}, \quad (5.1)$$

using the barred form of a bilocal operator $\mathcal{O}_{\rho\sigma}(x, y)$ that sums up all the tensors of higher and higher rank in Eq. (3.3).

Now in terms of bilocal operators we can formulate a much stronger hypothesis than the modified Wilson hypothesis mentioned in the last section. There we supposed that on the right-hand side of any light-cone commutators (unless the leading terms were forbidden for some reason) we would always have operators with $l = -J - 2$ and that for a given J and a given set of quantum numbers there would be very few of these, perhaps only one, and that the quantum numbers themselves would be greatly restricted (for example, to SU_3 octets and singlets). Here we can state the much stronger conjecture that for a given set of quantum numbers the bilocal operators appearing on the right are very few in number (and perhaps there is only one in each case), with the quantum numbers greatly restricted. That means that instead of an arbitrary series $\mathcal{O}_{\rho\sigma}$ + const. $z_\lambda z_\mu \mathcal{O}_{\rho\sigma\lambda\mu}$ + const.$' z_\lambda z_\mu z_\alpha z_\beta \mathcal{O}_{\alpha\beta\rho\sigma\lambda\mu} + \cdots$, we have a unique sum $\mathcal{O}_{\rho\sigma}(x, y)$ with all the constants determined. The same bilocal operator will appear in many commutators, then, and its matrix elements (for example, between proton and proton with no charge of momentum) will give universal deep inelastic form factors.

Let us express in terms of bilocal operators the idea mentioned in the last section that all tensor operators appearing on the right-hand side of the light cone current commutators may themselves be commuted according to conservation of dimension on the light cone, but lead to the same set of operators, giving a closed light cone algebra of an infinite number of local operators of all tensor ranks. We can sum up all these operators to make bilocal operators and commute those, obtaining, on the right-hand side according to the principle mentioned above, the same bilocal operators. Thus we obtain a light cone algebra generated by a small finite number of bilocal operators. These are the bilocal operators that give the most singular terms on the light cone in any commutator of local operators, the terms that give scaling behavior. (As we have said, in certain cases they may be forbidden to occur and positive powers of masses would then appear instead of dimensionless coefficients.)

This idea of a universal light cone algebra of bilocal operators with $l = -J - 2$ is a very elegant hypothesis, but one that goes far beyond present experimental evidence. We can hope to check it some day if we can find situations in which limiting cases of experiments involve the light cone commutators of light cone commutators. Attempts have been made to connect differential cross sections for the Compton effect with such mathematical quantities;[5] it will be interesting to see what comes of that and other such efforts.

A very important technical question arises in connection with the light cone algebra of bilocal operators. When we talk about the commutators of the individual local operators of all tensor ranks, we are dealing with just two points x and y and with the limit $(x - y)^2 \to 0$. But when we treat the commutator of bilocal operators $\mathcal{O}(x, u)$ and $\mathcal{O}(y, v)$, what are the space-time relationships of x, u, y, and v in the case to which the commutation relations apply? We must be careful, because if we give too liberal a prescription for these relationships we may be assuming more than could be true in any realistic picture of hadrons.

The bilocal operators arise originally in commutators of local operators on the light cone, and therefore we are interested in them when $(x - u)^2 \to 0$ and $(y - v)^2 \to 0$. In the light cone algebra of bilocal operators, we are interested in singularities that are picked up when $(x - y)^2$ or when $(u - v)^2 \to 0$ or when $(x - v)^2 \to 0$ or when $(u - y)^2 \to 0$. But do we have to have all six quantities simultaneously brought near to zero? That is not yet clear. In order to be safe, let us assume here that all six quantities do go to zero.

6. LIGHT CONE ALGEBRA ABSTRACTED FROM A QUARK PICTURE

Can we postulate a particular form for the light cone algebra of bilocal operators?

We have indicated above that if the Stanford experiments, when extended and refined, still suggest the absence of logarithmic terms the vanishing of the longitudinal cross section, and a difference between neutron and proton in the deep inelastic limit, then it looks as if in this limit Nature is following free field theory, or interacting field theory with naïve manipulation of operators, rather than what we know about the perturbation expansions of renormalised field theory. We might, therefore, look at a simple relativistic field theory model and abstract from it a light cone algebra that we could postulate as being true of the real system of hadrons. The simplest such model would be that of free quarks.

In the same way, the idea of an algebra of equal-time commutators of charges or charge densities was abstracted ten years ago from a relativistic Lagrangian model of a free spin 1/2 triplet, what would nowadays be called the quark triplet. The essential feature in this abstraction was the remark that turning on certain kinds of strong interaction in such a model would not affect the equal time commutation relations, even when all orders of perturbation theory were included; likewise, mass differences breaking the symmetry under SU_3 would not disturb the equal time commutation relations of SU_3.

We are faced, then, with the following question. Are there non-trivial field theory models of quarks with interactions such that the light cone algebra of free quarks remains undisturbed to all orders of naïve perturbation theory? Of course, the interactions will make great changes in the operator commutators inside the light cone; the question is whether the leading singularity on the light cone is unaffected. Let us assume, for purposes of our discussion, that the answer is affirmative. Then we can feel somewhat safe from absurdity in postulating for real hadrons the light cone algebras of free quarks, and indeed of massless free quarks (since the masses do not affect the light cone singularity).

Actually, it is easy to construct an example of an interacting field theory in which our condition seems to be fulfilled, namely a theory in which the quark field interacts with a neutral scalar or pseudoscalar "gluon" field ϕ. We note the fact that the only operator series in such a theory that fulfills $l = -J - 2$ and contains $\phi(x)$ is the following: $\phi(x)\phi(x)$, $\phi(x)\partial_\mu\phi(x)\cdots$. But these operators do not seem to appear in light cone expansions of products of local operators consisting only of quark fields, like the currents. A different situation prevails in a theory in which the "gluon" is a vector meson, since in that case we can have the operator series $\bar{q}(x)\gamma_\mu B_\nu(x)q(x)$, $\bar{q}(x)\gamma_\mu B_\nu B_\rho q(x)$, $\cdots$, contributing to the Bjorken limit. The detailed behavior of the various "gluon" models is being studied by Llewellyn Smith.[18]

In the following, we consider the light cone algebra suggested by the quark model. We obtain for the commutator of two currents on the light cone (connected part only):

$$
\begin{aligned}
&[\mathscr{F}_{i\mu}(x), \mathscr{F}_{j\nu}(y)] \\
&\quad \hat{=} \frac{1}{4\pi}\partial_\rho[\varepsilon(z_o)\delta(z^2)]\{if_{ijk}[s_{\mu\nu\rho\sigma}(\mathscr{F}_{k\sigma}(x,y) + \mathscr{F}_{k\sigma}(y,x)) \\
&\qquad + i\varepsilon_{\mu\nu\rho\sigma}(\mathscr{F}^5_{k\sigma}(y,x) - \mathscr{F}^5_{k\sigma}(x,y))] + d_{ijk}[s_{\mu\nu\rho\sigma}(\mathscr{F}_{k\sigma}(x,y) \\
&\qquad - \mathscr{F}_{k\sigma}(y,x)) - i\varepsilon_{\mu\nu\rho\sigma}(\mathscr{F}^5_{k\sigma}(y,x) + \mathscr{F}^5_{k\sigma}(x,y))]\},
\end{aligned}
$$

$$\begin{aligned}
&[\mathcal{F}^5_{i\mu}(x), \mathcal{F}_{j\nu}(y)] \\
&\quad \hat{=} \frac{1}{4\pi}\partial_\rho[\varepsilon(z_0)\delta(z^2)]\{if_{ijk}[s_{\mu\nu\rho\sigma}(\mathcal{F}^5_{k\sigma}(x,y) + \mathcal{F}^5_{k\sigma}(y,x)) \\
&\quad + i\varepsilon_{\mu\nu\rho\sigma}(\mathcal{F}_{k\sigma}(y,x) - \mathcal{F}_{k\sigma}(x,y))] \\
&\quad + d_{ijk}[s_{\mu\nu\rho\sigma}(\mathcal{F}^5_{k\sigma}(x,y) - \mathcal{F}^5_{k\sigma}(y,x)) \\
&\quad - i\varepsilon_{\mu\nu\rho\sigma}(\mathcal{F}_{k\sigma}(y,x) + \mathcal{F}_{k\sigma}(x,y))]\}, \\
&[\mathcal{F}^5_{i\mu}(x), \mathcal{F}^5_{j\nu}(y)] = [\mathcal{F}_{i\mu}(x), \mathcal{F}_{j\nu}(y)], \\
&s_{\mu\nu\rho\sigma} = \delta_{\mu\rho}\delta_{\nu\sigma} + \delta_{\nu\rho}\delta_{\mu\sigma} - \delta_{\mu\nu}\delta_{\rho\sigma}, \qquad z = x - y.
\end{aligned} \tag{6.1}$$

If we go to the equal time limit in (6.1) we pick up the current algebra relations for the currents; in fact we obtain, for the space integrals of all componets of nine vector and nine axial-vector currents, the algebra[19] of $U_6 \times U_6$.

Note that we can get similar relations for the current anti-commutators or for the products of currents on the light cone, just by replacing

$$\frac{1}{4\pi}\partial_\rho[\varepsilon(z_0)\delta(z^2)] \text{ by } -\frac{i}{4\pi^2}\partial_\rho\frac{1}{z^2} \text{ or by } -\frac{i}{8\pi^2}\partial_\rho\frac{1}{z^2 + i\varepsilon z_0}$$

respectively. Perhaps we can abstract these relations also and use them for hadron theory.

In (6.1) we have introduced bilocal generalizations of the vector and axial-vector currents, which in a quark model correspond to products of quark fields:

$$\begin{aligned}
\mathcal{F}_{k\sigma}(x,y) &\sim \bar{q}(x)\frac{i}{2}\lambda_k\gamma_\sigma q(y), \\
\mathcal{F}^5_{k\sigma}(x,y) &\sim \bar{q}(x)\frac{i}{2}\lambda_k\gamma_\sigma\gamma_5 q(y).
\end{aligned} \tag{6.2}$$

Note that the products in (6.2) have to be understood as "generalized Wick products". The c-number part in the product of two quark fields is already excluded, since it does not contribute to the connected current commutator. The c-number part is measured by vacuum processes like e^+e^- annihilation. Assuming that the disconnected part of the commutator on the light cone is also dictated by the quark model, we would obtain

$$\sigma_{\text{tot } e^+e^-} \sim \text{const.}/s \text{ for } e^+e^- \text{ annihilation,}$$

where s is as usually defined: $s = -(p_1 + p_2)^2$. In particular, we would get

$$\sigma_{\text{tot}}(e^+ e^- \text{ into hadrons}) \to (\Sigma Q^2)\sigma_{\text{tot}}(e^+ e^- \text{ into muons})$$

with $\Sigma Q^2 = (2/3)^2 + (1/^3)^2 + (1/3)^2 = 2/3$.

Now we go on to close the algebraic system of (6.1), where local currents occur on the left-hand side and bilocal ones on the right.

Let us assume that the bilocal generalizations of the vector and axial vector currents are the basic entities of the scheme. Again using the quark model as a guideline on the light cone, we obtain the following closed algebraic system for these bilocal operators:

$$\begin{aligned}
&[\mathscr{F}_{i\mu}(x,u), \mathscr{F}_{j\nu}(y,v)] \\
&\quad \hat{=} \frac{1}{4\pi}\partial_\rho\{\varepsilon(x_0 - v_0)\delta[(x-v)^2]\}(if_{ijk} - d_{ijk})(s_{\mu\nu\rho\sigma}\mathscr{F}_{k\sigma}(y,u) \\
&\qquad + i\varepsilon_{\mu\nu\rho\sigma}\mathscr{F}^5_{k\sigma}(y,u)) \\
&\qquad + \frac{1}{4\pi}\partial_\rho\{\varepsilon(u_0 - y_0)\partial[(u-y)^2]\}(if_{ijk} + d_{ijk}) \\
&\qquad \cdot (s_{\mu\nu\rho\sigma}\mathscr{F}_{k\sigma}(x,v) - i\varepsilon_{\mu\nu\rho\sigma}\mathscr{F}^5_{k\sigma}(x,v)), \\
&[\mathscr{F}^5_{i\mu}(x,u), \mathscr{F}_{j\nu}(y,v)] \qquad\qquad (6.3) \\
&\quad \hat{=} \frac{1}{4\pi}\partial_\rho\{\varepsilon(x_0 - v_0)\delta[(x-v)^2]\}(if_{ijk} - d_{ijk}) \\
&\qquad (s_{\mu\nu\rho\sigma}\mathscr{F}^5_{k\sigma}(y,u) + i\varepsilon_{\mu\nu\rho\sigma}\mathscr{F}_{k\sigma}(y,u)) \\
&\qquad + \frac{1}{4\pi}\partial_\rho\{\varepsilon(u_o - y_o)\delta[(u-y)^2]\}(if_{jik} + d_{ijk}) \\
&\qquad \cdot (s_{\mu\nu\rho\sigma}\mathscr{F}^5_{k\sigma}(x,v) - i\varepsilon_{\mu\nu\rho\sigma}\mathscr{F}_{k\sigma}(x,v)), \\
&[\mathscr{F}^5_{i\mu}(x,u), \mathscr{F}^5_{j\nu}(y,v)] \hat{=} [\mathscr{F}_{i\mu}(x,u), \mathscr{F}_{j\nu}(y,v)]
\end{aligned}$$

Similar relations might be abstracted for the anticommutators and products of two bilocal currents near the light cone. The relations (6.3) are assumed to be true if

$$(x-u)^2 \approx 0, \quad (u-y)^2 \approx 0,$$
$$(u-v)^2 \approx 0, \quad (x-y)^2 \approx 0,$$
$$(x-v)^2 \approx 0, \quad (u-v)^2 \approx 0,$$

This condition is obviously fulfilled if the four points x, u, y, v are distributed on a straight line on the light cone. The algebraic relations (6.3) can be used, for example, to determine the light cone commutator of two light cone commutators and relate this more complicated case to the simpler case of a light cone commutator. It would be interesting to propose experiments in order to test the relations (6.3).

7. LIGHT CONE ALGEBRA AND DEEP INELASTIC SCATTERING

In the last section we have emphasized that perhaps the light cone is a region of very high symmetry (scale and $SU_3 \times SU_3$ invariance). Furthermore, we have abstracted from the quark model certain algebraic properties that might be right on the light cone. Now we should like to mention some general relations that we can obtain using this light cone algebra. But let us first consider the weak interactions in the deep inelastic region.

We introduce the weak currents $J_\mu^+(x)$, $J_\nu^-(x)$ and consider the following expression:

$$W_{\mu\nu}(q) = \frac{1}{4\pi}\int d^4z e^{-iq\cdot z}\langle p|[J_\mu^+(z), J_\nu^-(0)]|p\rangle$$
$$= \left(\delta_{\mu\nu} - \frac{q_\mu q_\nu}{q^2}\right)\left(W_1^+ - \frac{(p\cdot q)^2}{q^2}W_2^+\right) - \frac{i}{2}\epsilon_{\mu\nu\alpha\beta}p_\alpha q_\beta W_3^+$$
$$+ \frac{\delta_{\mu\nu}(p\cdot q)^2 + p_\mu p_\nu q^2 - (p_\mu q_\nu + p_\nu q_\mu)p\cdot q}{q^2}W_2^+ + q^\mu q^\nu W_4^+$$
$$+ (q_\mu p_\nu + q_\nu p_\mu)W_5^+ + i(q_\mu p_\nu - q_\nu p_\mu)W_6^+ . \tag{7.1}$$

In general, we have to describe the inelastic neutrino hadron processes by six structure functions. From naïve scaling arguments we would expect in the deep inelastic limit:

$$W_1{}^+ \to F_1(\xi), \qquad -q\cdot pW_2{}^+ \to F_2(\xi),$$
$$-q\cdot pW_3{}^+ \to F_3(\xi), \qquad -q\cdot pW_4{}^+ \to F_4(\xi), \qquad (7.2)$$
$$-q\cdot pW_5{}^+ \to F_5(\xi), \qquad -q\cdot pW_6{}^+ \to F_6(\xi).$$

The formulae above have the most general form, valid for arbitrary vectors $J_\mu(x)$. We neglect the T-violating effects, which may in any case be 0 on the light cone: $F_6 = 0$. We have already stressed that the weak currents are conserved on the light cone, and we conclude:

$$F_4(\xi) = F_5(\xi) = 0. \qquad (7.3)$$

Equation (7.3) is an experimental consequence of the $SU_3 \times SU_3$ symmetry on the light cone, which may be tested by experiment. In the deep inelastic limit we have only three non-vanishing structure functions, corresponding to a conserved current.

It is interesting to note that there is the possibility of testing the dimension l of the divergence of the axial vector current, if our scaling hypothesis is right. We write, for the weak axial vector current,

$$\partial_\mu \mathscr{F}^5_{\pm\mu} = c \cdot v_\pm(x) \qquad (7.4)$$

where $v_\pm(x)$ is a local operator of dimension l, and c is a parameter with non-zero dimension.

According to our assumptions about symmetry breaking, c can be written as a positive power of a mass. Using (7.1), we obtain

$$\begin{aligned} q^\mu q^\nu W_{\mu\nu}{}^+(q) &= \frac{c^2}{4\pi}\int d^4z e^{-iq\cdot z}\langle p|[v_+(z), v_-(0)]|p\rangle \\ &= (q^2)^2 W_4{}^+ - 2q^2 q\cdot pW_5{}^+. \end{aligned} \qquad (7.5)$$

We define:

$$D(q^2, q\cdot p) = \frac{1}{4\pi}\int d^4z e^{-iq\cdot z}\langle p|[v_+(z), v_-(0)]|p\rangle. \qquad (7.6)$$

If we assume that D scales in the deep inelastic region according to the dimension l of $v_\pm(x)$, we obtain

$$\lim_{bj} (-p\cdot q)^{-l-3} D(q^2, q\cdot p) = \phi(\xi) \qquad (7.7)$$

where $\phi(\xi)$ denotes the deep inealstic structure function for the matrix element (7.6).

Using (7.5) we obtain

$$\lim_{bj} (-p \cdot q)^{5+l} (\xi^2 W_4{}^+ - 2\xi W_5{}^+) = c^2 \phi(\xi). \tag{7.8}$$

If we determine experimentally the scaling properties of W_4 and W_5, then we can deduce from (7.8) the dimension l of $v_{\pm}(x)$. This l is the same quantity as the dimension l_u discussed in Section 2, provided the $SU_3 \times SU_3$ violating term in the energy has a definite dimension.[20]

In order to apply the light cone algebra of Section 6, we have to relate the expectation values of the bilocal operators appearing there to the structure function in question. This is done in Appendix II, where we give this connection for arbitrary currents. We use Eqs. (A.12) and (A.13), where the functions $S^k(\xi)$, $A^k(\xi)$ are given by the expectation value of the symmetric and antisymmetric bilocal currents (Eq. (A.8)), and obtain:

(a) for deep inelastic electron-hadron scattering:

$$F_2^{ep}(\xi) = \xi\left(\frac{2}{3}\sqrt{\frac{2}{3}} A^0(\xi) + \frac{1}{3\sqrt{3}} A^8(\xi) + \frac{1}{3} A^3(\xi)\right) \tag{7.9}$$

(b) for deep inelastic neutrino-hadron scattering:

$$F_2^{\nu p}(\xi) = \xi(2S^3(\xi) + 2\sqrt{\frac{2}{3}} A^0(\xi) + \frac{2}{\sqrt{3}} A^8(\xi)) \tag{7.10}$$

$$F_3{}^{\nu p}(\xi) = 2A^3(\xi) - 2\sqrt{\frac{2}{3}} S^0(\xi) - \frac{2}{\sqrt{3}} S^8(\xi). \tag{7.11}$$

In (7.5) and (7.6) we have neglected the Cabibbo angle, since $\sin^2\theta_c = 0.05 \approx 0$.

Both in (7.4) and (7.6), $A^3(\xi)$ occurs as the only isospin dependent part, and we can simply derive relations between the structure functions of different members of an isospin multiplet, e.g., for neutron and proton:

$$6 \cdot (F_2^{en} - F_2^{ep}) = \xi \cdot (F_3^{\nu p} - F_3^{\nu n}). \tag{7.12}$$

This relation was first obtained by C. H. Llewellyn Smith[7] within the "parton" model. One can derive similar relations for other isospin multiplets.

In the symmetric bilocal current appear certain operators that we know. The operator $j_\mu(x) = i\bar{q}(x)\gamma_\mu q(x)$ has to be identical with the hadron current (we suppress internal indices) in order to give current algebra. But we know their expectation values, which are given by the corresponding quantum number. In such a way we can derive a large set of sum rules relating certain moments of the structure functions to their well-known expectation values.

We give only the following two examples, which follow immediately from (7.9), (7.10), (7.11):

$$\int_{-1}^{1} \frac{d\xi}{\xi}(F_2^{\nu p}(\xi) - F_2^{\nu n}(\xi)) = \int_{-1}^{1}\frac{d\xi}{\xi}(F_2^{\nu p}(\xi) - F_2^{\nu p}(-\xi)) = 4s_1^3(p) = 4. \tag{7.13}$$

Here $s_1^3(p)$ means, as in Appendix II, the proton expectation value of $2F_3$. This is the Adler sum rule,[21] usually written as

$$\int_0^1 \frac{d\xi}{\xi}(F_2^{\nu p}(\xi) - F_2^{\nu n}(\xi)) = 2. \tag{7.14}$$

From (7.11) we obtain:

$$\int_1^1 (F_3^{\nu p} + F_3^{\nu n})d\xi = -2(2s_1^0(p) + s_1^8(p)) = -12 \tag{7.15}$$

or

$$\int_{-0}^{1} (F_3^{\nu p} + F_3^{\nu n})d\xi = -6, \tag{7.16}$$

which is the sum rule first derived by Gross and Llewellyn Smith.[22]

If we make the special assumption that we are abstracting our light cone relations from a pure quark model with no "gluon field" and non-derivative couplings, we can get a further set of relations. Of course, no such model is known to exist in four dimensions that is even renormalizable, much less super-renormalisable as we would prefer to fit in with the ideas presented here. Nevertheless, it may be worthwhile to examine sum rules that test whether Nature imitates the "pure quark" case.

The point is that when we expand the bilocal quantity $\mathcal{F}_{0\alpha}(x, y)$ to first order in $y - x$, we pick up a Lorentz tensor operator, a singlet under SU_3, that corresponds in the quark picture to the operator $1/2\{\bar{q}(x)\gamma_\mu\partial_\nu q(x) - \partial_\nu\bar{q}(x)\gamma_\mu q(x)\}$, which, if we symmetrize in μ and ν and ignore the trace, is the same as the stress-energy-momentum tensor $\theta_{\mu\nu}$ in the pure quark picture. But the expected value of $\theta_{\mu\nu}$ in any state of momentum p is just $2p_\mu p_\nu$ and so we obtain sum rules for the pure quark case.

We consider the isospin averaged expressions:

$$(F_2^{ep}(\xi) + F_2^{en}(\xi)) = 2\xi\left\{\frac{2}{3}\sqrt{\frac{2}{3}}A^0(\xi) + \frac{1}{3}\frac{1}{\sqrt{3}}A^8(\xi)\right\}$$

$$(F_2^{\nu p}(\xi) + F_2^{\nu n}(\xi)) = 2\xi\left\{2\sqrt{\frac{2}{3}}A^0(\xi) + \frac{2}{\sqrt{3}}A^8(\zeta)\right\}$$

and obtain

$$6(F_2^{ep} + F_2^{en}) - (F_2^{\nu p} + F_2^{\nu n}) = 4\sqrt{\frac{2}{3}}A^0(\xi)$$

$$= 4\sqrt{\frac{2}{3}}(a_1^0\delta(\xi) - \frac{1}{2!}a_3^0\delta''(\xi))\cdots)$$

In pure quark theories we have $a_1^0 = \sqrt{2/3}$ and we obtain

$$6\int_{-1}^{1}(F_2^{ep} + |F_2^{en})d\xi - \int_{-1}^{1}(F_2^{\nu p}(\xi) + F_2^{\nu n}(\xi))d\xi = 8/3$$

or, for the physical region $0 \leqq \xi \leqq 1$:

$$6\int_0^1(F_2^{ep} + F_2^{en})d\xi - \int_0^1(F_2^{\nu p} + F_2^{\nu n})d\xi = 4/3. \quad (7.17)$$

The sum role (7.17) can be tested by experiment. This will test whether one can describe the real world of hadrons by a theory resembling one with only quarks, interacting in some unknown non-linear fashion.

The scaling behavior in the deep inelastic region may be described by the "parton model".[4,5] In the deep inelastic region, the electron is viewed as scattering in the impulse approximation off point-like constituents of the hadrons ("partons"). In this case the scaling function $F_2^e(\xi)$ can be written as

$$F_2^e(\xi) = \sum_N P(N) \, (\sum_i Q_i^2)_N \xi f_N(\xi) \tag{7.18}$$

where we sum up over all "partons" (Σ_i) and all the possibilities of having N partons (Σ_N). The momentum distribution function of the "partons" is denoted by $f_N(\xi)$, the charge of the i-th "parton" by Q_i. We compare (7.9) with (7.18):

$$\begin{aligned} F_2^e(\xi) &= \xi \, \frac{2}{3}A^0(\xi) + \frac{1}{6}A^8(\xi) + \frac{1}{3}A^3(\xi)) \\ &= \sum_N P(N) \left(\sum_i Q_i^2)_N \xi f_N(\xi). \right) \end{aligned} \tag{7.19}$$

As long as we do not specify the functions $f_N(\xi)$ and $P(N)$, the "parton model" gives us no more information than the generalization of current algebra to the light cone as described in the last sections. If one assumes special properties of these functions, one goes beyond the light cone algebra of the currents, that means beyond the properties of the operator products on the light cone. Such additional assumptions, e.g., statistical assumptions about the distributions of the "partons" in relativistic phase space, appear in the light cone algebra approach as specific assumptions about the matrix elements of the expansion operators on the light cone. These additional assumptions are seen, in our approach, to be model dependent and somewhat arbitrary, as compared to results of the light cone algebra. Our results can, of course, be obtained by "parton" methods and are mostly well-known in that connnection.

It is interesting to consider the different sum roles within the "parton model". The sum rules (7.14) and (7.16) are valid in any "quark-parton" model; so is the symmetry relation (7.12). The sum rule (7.17) is a specific property of a model consisting only of quarks. If there is a "gluon" present, we obtain a deviation from 4/3 on the right-hand side, which measues the "gluon" contribution to the energy-momentum tensor.

Our closed algebra of bilocal operators on the light cone has, of course, a parallel in the "parton" model. However, it is again much easier using our approach to disentangle what may be exactly true (formulae for light cone commutators of light cone commutators) from what depends on specific matrix elements and is therefore model dependent. It would be profitable to apply such an analysis to the work of Bjorken and Paschos, in the context of "partons", on scaling in the Compton effect on protons.

As an example of a "parton model" relation that mingles specific assumptions about matrix elements with more general ideas of light cone algebra and abstraction from a pure quark model, we may take the allegation that in the pure quark case we have $\int F_2^{en}(\xi)d\xi = 2/9$. Light cone algebra and the pure quark assumption do not imply this.

8. CONCLUDING REMARKS

There are many observations that we would like to make and many unanswered questions that we would like to raise about light cone algebra. But we shall content ourselves with just a few remarks.

First comes the question of whether we can distinguish in a well-defined mathematical way, using physical quantities, between a theory that makes use of SU_3 triplet representations locally and one that does not. If we can, we must then ask whether a theory that has triplets locally necessarily implies the existence of real triplets (say real quarks) asymptotically. Dashen (private communication) raises these two questions by constructing local charge operators $\int_V \mathscr{F}_{i0} d^3x$ over a finite volume. (This construction is somewhat illegitimate, since test functions in field theory have to be multiplied by δ functions in equal time charge density commutators and should therefore have all derivatives, not like the function that Dashen uses, which is unity inside V and zero outside.) If his quantities F_i^V make sense, they obey the commutation rules of SU_3 and we can ask whether for any V our states contain triplet (or other triality $\neq 0$) representations of this SU_3. Dashen then suggests that our bilocal algebra probably implies that local triplets in this sense are present; if the procedure and the conclusion are correct, we must ask whether real quarks are then implied.

The question of quark statistics is another interesting one. If quarks are real, then we cannot assign them para- Fermi statistics of rank 3, since that is said to violate the factoring of the S-matrix for distant subsystems. However, if somehow our quarks are permanently bound in oscillators (and our theory is thus perhaps equivalent to a bootstrap theory with no real quarks), then they could be parafermions of rank 3. They can be bosons, too, if they are not real, but only if there is a spinless fermion (the "soul" of a baryon) that accompanies the three quarks in each baryon.

Another topic is the algebra of $U_6 \times U_6 \times O_3$ that is implied at equal times for the integrals of the current component and the angular momentum.[19] Is that algebra really correct or is it too strong an assumption? Should it be replaced at $P_z = \infty$ by only the "good-good" part of the algebra?

If we do have the full algebra, then the quark kinetic part of the energy density is uniquely defined as the part behaving like **(35, 1)** and **(1, 35)** with $L = 1$, i.e., like $\boldsymbol{\alpha} \cdot \boldsymbol{\nabla}$.

If we abstract relations from a pure quark picture without gradient couplings, then this quark kinetic part of $\theta_{\mu\nu}$ is all there is apart from the trace contribution. In that case, we have the equal time commutation relation for the whole energy operator:

$$\sum_{r=1}^{3} \sum_{i=1}^{8} \left[\int \mathscr{F}_{ir} d^3x, \left[\int \mathscr{F}_{ir} d^3x, P_0 \right] \right] = 16/3\, P_0 + \text{scale violating terms.}$$

This relation, in the pure quark case, can be looked at in another way. It is an equal time consequence of the relation

$$\theta_{\mu\nu} = \lim_{y \to x} \frac{3\pi^2}{32} \partial_\mu \partial_\nu \{ (z^2)^2 \mathscr{F}_{i\alpha}(x) \mathscr{F}_{i\alpha}(y) \} + \text{scale violating terms}$$

that holds when the singlet tensor term in the light cone expansion of $\mathscr{F}_{i\mu}(x)\mathscr{F}_{j\nu}(y)$ is just proportional to $\theta_{\mu\nu}$, as in the pure quark case. This relation is what, in the pure quark version of the light cone algebra (extended to light cone products), replaces the Sugawara[23] model, in which $\theta_{\mu\nu}$ is proportional to $\mathscr{F}_{i\mu}\mathscr{F}_{i\nu}$, with dimension -6. Our expression is much more civilized, having $l = -4$ as it should. A more general equal time commutator than the one above, also implied by the pure quark case, is the following:

$$\sum_{r=1}^{3} [\mathcal{F}_{ir}(x), \partial_0 \mathcal{F}_{ir}(y)] = 16i/3\, \theta_{00}\delta(x-y) + \text{scale breaking terms.}$$

Another important point that should be emphasized is that the $U_6 \times U_6$ algebra requires the inclusion of a ninth vector current $\mathcal{F}_{0\alpha}$ and a ninth axial vector current $\mathcal{F}^5_{0\alpha}$, and that the Latin index for SU_3 representation components in Appendix II has to run from 0 to 8. Now if the term in the energy density that breaks $SU_3 \times SU_3$ follows our usual conjecture and behaves like $-u_0 - cu_8$ with c near $-\sqrt{2}$ and if the chiral symmetry preserving but scale breaking term δ is just a constant, then as $u \to 0$ scale invariance and chiral invariance become good, but the mass formula for the pseudoscalar mesons indicates that we do not want $\partial_\alpha \mathcal{F}^5_{0\alpha}$ to be zero in that limit.[10] Yet $\mathcal{F}^5_{0\alpha}$ is supposed to be conserved on the light cone. Does this raise a problem for the idea of δ = const. or does it really raise the whole question of the relation of the light cone limit and the formal limit $u \to 0$, $\delta \to 0$?

If there are dilations, with $m^2 \to 0$ in the limit of scale invariance while other masses stay finite, how does that jibe with the light cone limit in which all masses act as if they go to zero? Presumably there is no contradiction here, but the situation should be explored further.

Finally, let us recall that in the specific application of scaling to deep inelastic scattering, the functions $F(\xi)$ connect up with two important parts of particle physics. As $\xi \to 0$, if we can interchange this limit with the Bjorken limit, we are dealing with fixed q^2 and with $p \cdot q \to \infty$ and the behavior of the F's comes directly from the Regge behavior of the corresponding exchanged channel. If $\alpha_P(0) = 1$, then $F_2^{ep}(\xi) + F_2^{en}(\xi)$ goes like a constant at $\xi = 0$, i.e., $\xi^{1-\alpha_P(0)}$, while $F_2^{ep}(\xi) - F_2^{en}(\xi)$ goes like $^{1-\alpha_\rho(0)}$, etc.

As $\xi \to 1$, as emphasized by Drell and Yan[8], there seems to be a connection between the dependence of $F(\xi)$ on $1 - \xi$ and the dependence of the elastic form factors of the nucleons on t at large t.

ACKNOWLEDGEMENTS

We would like to thank J. D. Bjorken, R. P. Feynman, and C. H. Llewellyn Smith for stimulating conversations about the relation of our work to previous work on "partons". One of us (H. F.) would like to express his gratitude to the DAAD, to SLAC, and to the AEC high energy physics group at Caltech for support.

The ninth Section, prepared for the Tel Aviv Conference, contains a number of points that have been elaborated between the Conference and the time of publication, especially matters concerned with "anomalies". For many enlightening discussions of these questions, we are deeply indebted to W. Baarden, and to the staff of the Theoretical Study Division of CERN.

REFERENCES

1. K. G. Wilson, *Phys. Rev.* **179**, 1499 (1969).
2. R. Brandt and G. Preparata, CERN preprint TH-1208.
3. Y. Frishman, Phys. Rev. Lett. 25, 966 (1970).
4. R. P. Feynman, *Proceedings of Third High Energy Collision Conference at State University of New York*, Stony Brook, Gordon and Breach, 1970.
5. J. D. Bjorken and E. A. Paschos, Phys. Rev. **185**, 1975 (1969).
6. P. Landshoff and J. C. Polkinghorne, Cambridge University DAMTP preprints (1970).
7. C. H. Llewellyn Smith, Nucl. Phys. **B17**, 277 (1970).
8. S. D. Drell and T. Yan, Phys. Rev. Lett. **24**, 181 (1970).
9. Note we use the metric $\delta_{\mu\nu} = (1, 1, 1, -1)$ and the covariant state normalization $\langle p's' | ps \rangle = (2\pi)^3\, 2p_0^s\, (p-p')\, \delta_{s_s'}$.
10. M. Gell-Mann, *Proceedings of Third Hawaii Topical Conference on Particle Physics*, Western Periodicals Co., Los Angeles, 1969.
11. F. von Hippel and J. K. Kim, *Phys. Rev.* **D1**, 151 (1970).
12. J. Ellis, Physics Lett. **33B**, 591 (1970).
13. R. F. Dashen and T. P. Cheng, Institute for Advanced Study preprint (1970).
14. J. Ellis, P. Weisz, and B. Zumino, *Phys. Lett.* **34B**, 91 (1971).
15. E. D. Bloom, G. Buschorn, R. L. Cottrell, D. H. Coward, H. DeStaebler, J. Drees, C. L. Jordan, G. Miller, L. Mo, H. Piel, R. E. Taylor, M. Breidenbach, W. R. Ditzler, J. I. Friedman, G. C. Hartmann, H. W. Kendall, and J. S. Poucher, Stanford Linear Accelerator Center preprint SLAC-PUB-796 (1970) (report presented at the XVth International Conference on High Energy Physics, Kiev, USSR, 1970).
16. H. Leutwyler and J. Stern, *Nucl. Phys.* **B20**, 77 (1970); R. Jackiw, R. Van Royen, and G. B. West, *Phys. Rev.* **D2**, 2473 (1970).
17. S. Ciccariello, R. Gatto, G. Sartori, and M. Tonin, *Phys. Lett.* **30B**, 546 (1969); G. Mack, *Phys. Rev. Lett.* **25**, 400 (1970). J. M. Cornwall and R. E. Norton, *Phys. Rev.* **177**, 2584 (1968) used a different approach to accomplish about the same result. Instead of expanding light cone commutators, they use equal time commutators with higher and higher time derivatives, sandwiched between states at infinite momentum. That amounts to roughly the same thing, and represents an alternative approach to light cone algebra.
18. C. H. Llewellyn Smith, to be published.
19. R. P. Feynman, M. Gell-Mann, and G. Zweig, *Phys. Rev. Lett.* **13**, 678 (1964). The idea was applied to many important effects by J. D. Bjorken, Phys. Rev. **148**, 1467 (1966).
20. J. Mandula, A. Schwimmer, J. Weyers, and G. Zweig have proposed independently this test of the dimension l_μ and are publishing a full account of it.
21. S. Adler, *Phys. Rev.* **143**, 154 (1966).
22. D. J. Gross and C. H. Llewellyn Smith, *Nucl. Phys.* **B14**, 337 (1969).
23. H. Sugawara, *Phys. Rev.* **170**, 1659 (1968).
24. J. M. Cornwall and R. Jackiw, UCLA preprint (1971).
25. D. J. Gross and S. B. Treiman, Princeton University preprint (1971).
26. W. Bardeen, private communication.
27. M. Gell-Mann and F. E. Low *Phys. Rev.* **95**, 1300 (1954); M. Baker and K. Jonson, *Phys. Rev.* **183**, 1292 (1969).
28. A. H. Muller, *Phys. Rev.* **D2**, 2963 (1970).
29. J. D. Bjorken, Talk given at the same Conference.

Quantum Chromodynamics

Harald Fritzsch

A problem of the quark model was connected with the Pauli principle for the wave functions of the baryons, e.g. for the Ω-baryon. To solve this problem, William Bardeen, Fritzsch and Gell-Mann introduced in 1972 a new threefold quantum number for the quarks, which they described as the "colors" of the quarks.

There are red quarks, green quarks and blue quarks. The transformations of the colors are described by a color group SU(3). The three colors are denoted by "r", "g" and "b". The hadrons are assumed to be color singlets—thus a baryon wave function is a superposition of six terms:

$$baryon \Rightarrow (rgb - rbg + gbr - grb + brg - bgr).$$

The Ω-baryon can now be described as a function of the three colored strange quarks:

$$\Omega \sim \sum \varepsilon_{ikl} s^i s^k s^l,$$
$$i, k, l \sim r, g, b.$$

Due to the color quantum number the wave function is now antisymmetric, if two strange quarks are interchanged, and there is no problem with the Pauli principle.

H. Fritzsch (✉)
Physik-Department, Ludwig-Maximilians-Universität Physik-Department, München, Germany
e-mail: fritzsch@mppmu.mpg.de

H. Fritzsch (ed.), *Murray Gell-Mann and the Physics of Quarks*, Classic Texts in the Sciences, https://doi.org/10.1007/978-3-319-92195-2_7

All baryons are bound states of three quarks with different colors—here as an example the proton:

The meson wave function is the sum of three terms, involving the red, green and blue quarks:

$$meson \Rightarrow \left(\bar{r}r + \bar{g}g + \bar{b}b\right).$$

Another problem of the quark model is the decay rate of the neutral pion, which is an order of magnitude smaller as predicted. If the color quantum number is taken into account, the decay amplitude is increased by a factor 3, since in the triangle diagram the three colors of the quarks appear. Thus the decay rate is a factor 9 larger and agrees now with the experiment.

If an electron and a positron annihilate at high energies, a quark and an anti-quark are produced. We consider the ratio R, which is the ratio of the cross section to produce a quark pair and the cross section to produce a muon pair. This ratio is given by the sum of the squares of the charges of the quarks:

$$\mathrm{R} = (2/3)^2 + (-1/3)^2 + (-1/3)^2 = 2/3.$$

This ratio describes the production of hadrons, which are produced by the quark pair. According to the experiments the ratio R is about 2 in the energy region of about 2 GeV.

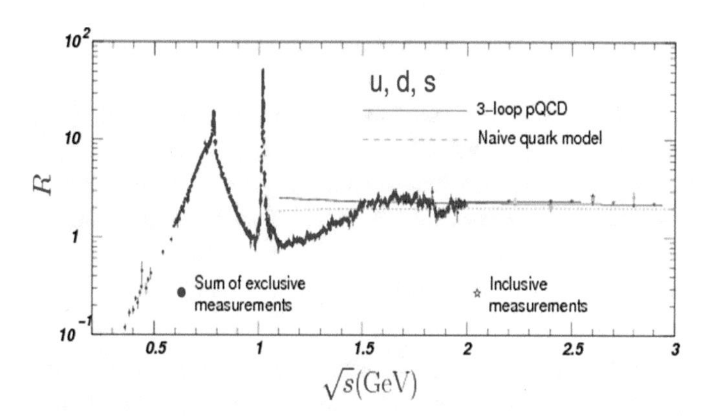

If the color quantum number is taken into account, the ratio R increases by a factor 3 and agrees with the experimental result. Thus there are three indications that the idea of the color quantum number is correct—the wave functions of the baryons, the decay of the neutral pion and the ratio R.

In 1972 Fritzsch and Gell-Mann started to investigate a gauge theory, using the colors of the quarks. The gauge group is the group SU(3) of the color transformations. This theory, which they called "Quantum Chromodynamics", is very similar to Quantum Electrodynamics, which describes the interaction of electrons with the electromagnetic field. It combines electrodynamics, quantum mechanics and the theory of relativity.

Quantum Electrodynamics has an interesting property. If the field of the electron is multiplied with a phase parameter, which depends on space and time, nothing changes. This is called "local gauge invariance". It requires that the electrons are interacting with a vector field, the field of the photon. Without this interaction the theory would not be invariant under local gauge transformations.

Furthermore the particle of the gauge field cannot have a mass—photons must be massless. A mass term would destroy local gauge invariance. The gauge transformations form a group, which in case of Quantum Electrodynamics is the unitary group U(1), the group of complex numbers with absolute value 1.

The strength of the interaction is described by the electromagnetic coupling constant e or by the fine-structure constant α:

$$\alpha = \frac{e^2}{4\pi}.$$

The fine-structure constant cannot be calculated. The experiments give:

$$\alpha \cong \frac{1}{137},$$
$$e \cong 03028.$$

In quantum electrodynamics the coupling parameter is not constant, but depends on the energy. Thus the fine-structure constant is not a constant, but a function of the energy. The value, given above, describes the fine-structure constant at zero energy. At increasing energy it increases slowly. At the mass of the Z-boson, about 91 GeV, it is about 6% larger:

$$\alpha(M_Z) \cong \frac{1}{129}.$$

In the theory of Quantum Chromodynamics the gauge group is the color group SU(3). The gauge bosons are the "gluons". The adjoint representation of the gauge group determines the number of gauge bosons. Here the adjoint representation is an octet, thus there are eight massless gluons.

In QCD the strength of the interaction is described by the coupling parameter g or the analogue of the fine-structure constant:

$$\alpha_s = \frac{g^2}{4\pi}.$$

In QED the gauge boson, the photon, interacts with the electron, but not with itself. In QCD this is not the case. The gluons interact with the quarks, but also with other gluons. This self-interaction of the gluons leads to an interesting property of QCD—asymptotic freedom. The gauge coupling parameter of QCD decreases, if the energy is increased. At very high energies the coupling parameter is small. It can be written as a function of an energy scale Λ:

$$\alpha_s\left(Q^2\right) = 2\pi/b \ln\left(Q^2/\Lambda^2\right),$$
$$b = 11 - \frac{2}{3}n.$$

Here n is the number of relevant quarks at the corresponding energy. In the energy region between 10 GeV and 300 GeV five quarks contribute, thus the parameter b is about 7.7.

Due to the asymptotic freedom the commutator of a current near light-like distances is very similar to the commutator in the free quark model. This implies that the quarks can be observed in deep inelastic scattering as nearly point-like constituents. At low energies the coupling constant might increase without limit, thus the quarks and gluons are confined. A rigorous proof of the confinement property is still missing.

In QCD the scaling property of the cross sections, observed in deep inelastic scattering, is not an exact property, but it is violated by small logarithmic terms. These scaling violations were observed and are in good agreement with the theoretical predictions.

The scaling violations are functions of the scale parameter Λ. The experiments are in agreement with the theoretical predictions, if this parameter is in the range

$$\Lambda = 213^{+38}_{-35}\ MeV.$$

Many experiments were carried out to measure the QCD coupling parameter:

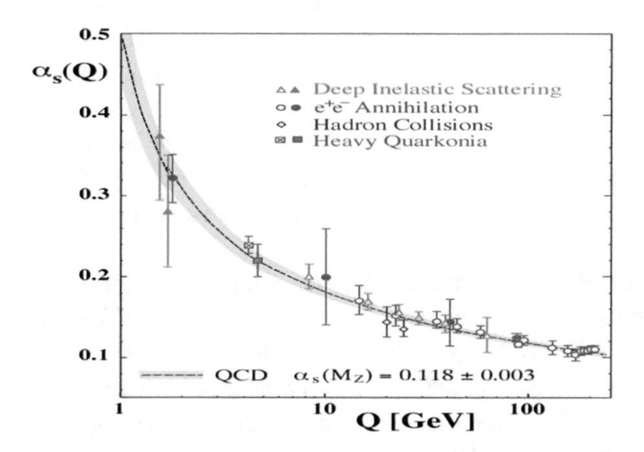

With the LEP accelerator at CERN one has determined the QCD coupling parameter at the mass of the Z-boson:

$$\alpha_s(M_Z) = 0.1184 \pm 0.0007.$$

In the absence of the quark masses the theory of QCD depends only on this scale parameter, which determines the properties of the hadrons, e.g. their masses or their magnetic moments. The proton mass can be expressed as a numerical constant, which can be calculated, multiplied by the scale parameter:

$$M_p \cong const. \cdot \Lambda.$$

Of course, in reality the proton mass also depends on the quark masses. About 20 MeV of the proton mass are due to the two u-quarks, about 19 MeV due to the d-quark, and about 35 MeV are due to the pairs of strange quarks and anti-quarks. If the quark masses are set to zero, the mass of the proton will be reduced to about 860 MeV.

Since the QCD coupling parameter increases, if the energy decreases, the force among the quarks becomes strong at large distances. Perturbation theory is useless in this region. But if the space-time is described by a lattice, one can get information about the force between the quarks at large distances using computers.

We consider the case of one heavy quark. In QCD the forces between this quark and its anti-quark does not decrease at large distances, but remains constant. Due to the self-interaction of the gluons the gluon field lines attract each other. The quark and its anti-quark are connected by a string of parallel gluon field lines and the force is constant.

In reality there are quarks with very small masses. If the heavy quarks are moved away from each other, the gluons produce pairs of virtual light quarks and anti-quarks, which produce mostly pions. Thus no string of gluon field lines is generated. The heavy quark produces together with a light anti-quark a heavy meson, e.g. a B-meson.

In electron-positron annihilation at high energy the virtual photon produces a quark and an anti-quark, which move away from each other with essentially the speed of light. These quarks produce many mesons, which move away in the same direction as the original quark. Thus a jet of mesons is produced—a quark jet. These jets were predicted in the year 1974 by Richard Feynman.

The quark jets were observed at DESY in 1978. One year later one observed at DESY events with three jets. In the annihilation of an electron and a positron a quark, an anti-quark and a gluon were produced. The gluon produced also a particle jet, thus three jets were observed.

Fritzsch and Gell-Mann discussed in 1972 in their paper for the proceedings of the Rochester conference in Chicago neutral particles, composed of gluons. Since the gluons are color octets, two gluons could form a color singlet hadron, a glue-meson. One has searched for these mesons, but nothing was observed thus far. Presumably these mesons are difficult to observe—they have a high mass, mix with mesons, composed of a quark and an anti-quark, and are very unstable.

Besides the glue mesons there should exist also new baryons, composed of four quarks and an anti-quark, the penta-quark baryons. Thus far these particles have not been clearly observed. The LHCb collaboration at CERN reported the observation of a penta-quark baryon, composed of two u-quarks, one d-quark, one charmed quark and one charmed anti-quark. The mass of this particle is about 4.4 GeV.

There should exist also new mesons, composed of two quarks and two anti-quarks, the tetra-quark mesons. They should have masses of about 4 GeV. Thus far no clear evidence for the existence of these mesons has been found.

Shortly after the Big Bang the universe was filled with a hot, dense soup of particles. This mixture was dominated by quarks and by gluons. In those first moments of extreme temperature, however, quarks and gluons were bound only weakly, free to move on their own in a quark-gluon plasma.

To recreate conditions similar to those of the very early universe, accelerators produce head-on collisions between massive ions, such as gold or lead nuclei. In these heavy-ion collisions the hundreds of protons and neutrons in the two nuclei collide. This creates a quark-gluon plasma, which is in particular studied with the ALICE experiment at the Large Hadron Collider in CERN.

In heavy-ion collisions the quark-gluon plasma exists only for a very short time. But in the center of large neutron stars should be a quark-gluon plasma, which exists for a very long time.

Pion Decay and Electron-Positron Annihilation

Harald Fritzsch

H. Fritzsch (✉)
Physik-Department, Ludwig-Maximilians-Universität Physik-Department, München, Germany
e-mail: fritzsch@mppmu.mpg.de

H. Fritzsch (ed.), *Murray Gell-Mann and the Physics of Quarks*, Classic Texts in the Sciences, https://doi.org/10.1007/978-3-319-92195-2_8

CHAPTER 7

Light-Cone Current Algebra, π° Decay, and $e^{+}e^{-}$ Annihilation

W. A. BARDEEN
H. FRITZSCH
M. GELL-MANN

1. INTRODUCTION

The indication from deep inelastic electron scattering experiments at SLAC that Bjorken scaling may really hold has motivated an extension of the hypotheses of current algebra to what may be called light-cone current algebra.[1] As before, one starts from a field theoretical quark model (say one with neutral vector "gluons") and abstracts exact algebraic results, postulating their validity for the real world of hadrons. In light-cone algebra, we abstract the most singular term near the light cone in the commutator of two-vector or axial vector currents, which turns out to be given in terms of bilocal current operators that reduce to local currents when the two space-time points coincide. The algebraic properties of these bilocal operators, as abstracted from the model, give a number of predictions for the Bjorken functions in deep inelastic electron and neutrino experiments. None is in disagreement with experiment. These algebraic properties, by the way, are the same as in the free quark model.

Reprinted from *Scale and Conformal Symmetry in Hadron Physics*, ed. R. Gatto (© John Wiley & Sons, Inc., 1973), pp. 139–151.

From the mathematical point of view, the new abstractions differ from the older ones of current algebra (commutators of "good components" of current densities at equal times or on a light plane) in being true only formally in a model with interactions, while failing to each order of renormalized perturbation theory, like the scaling itself. Obviously it is hoped that, if the scaling works in the real world, so do the relations of light-cone current algebra, in spite of the lack of cooperation from renormalized perturbation theory in the model.

The applications to deep inelastic scattering involve assumptions only about the connected part of each current commutator. We may ask whether the disconnected part—for example, the vacuum expected value of the commutator of currents—also behaves in the light-cone limit as it does formally in the quark-gluon model, namely, the same as for a free quark model. Does the commutator of two currents, sandwiched between the hadron vacuum state and itself, act at high momenta exactly as it would for free quark theory? If so, then we can predict immediately and trivially the high-energy limit of the ratio

$$\sigma(e^+ + e^- \to \text{hadrons})/\sigma(e^+ + e^- \to \mu^+ + \mu^-)$$

for one-photon annihilation.

In contrast to the situation for the connected part and deep inelastic scattering, the annihilation results depend on the statistics of the quarks in the model. For three Fermi-Dirac quarks, the ratio would be $(\frac{2}{3})^2 + (-\frac{1}{3})^2 + (-\frac{1}{3})^2 = \frac{2}{3}$, but do we want Fermi-Dirac quarks? The relativistic "current quarks" in the model, which are essentially square roots of currents, are of course not identical with "constituent quarks" of the naive, approximate quark picture of baryon and meson spectra. Nevertheless, there should be a transformation, perhaps even a unitary transformation, linking constituent quarks and current quarks (in a more abstract language, a transformation connecting the symmetry group $[SU(3) \times SU(3)]_{W,\infty,\text{strong}}$ of the constituent quark picture of baryons and mesons, a subgroup of $[SU(6)]_{W,\infty,\text{strong}}$,[2] with the symmetry group $[SU(3) \times SU(3)]_{W,\infty,\text{currents}}$,[3] generated by the vector and axial vector charges). This transformation should certainly preserve quark statistics. Therefore the indications from the constituent quark picture that quarks obey peculiar statistics should suggest the same behavior for the current quarks in the underlying relativistic model from which we abstract the vacuum behavior of the light-cone current commutator.[4]

In the constituent quark picture of baryons,[5] the ground-state wave function is described by $(\mathbf{56},\mathbf{1}), L=0^+$ with respect to $[SU(6)\times SU(6)\times SU(3)]$ or $(\mathbf{56}, L_z=0)$ with respect to $[SU(6)\times O(2)]_W$. It is totally symmetric in spin and SU(3). In accordance with the simplicity of the picture, one might expect the space wave function of the ground state to be totally symmetric. The entire wave function is then symmetrical. Yet baryons are to be antisymmetrized with respect to one another, since they do obey the Pauli principle. Thus the peculiar statistics suggested for quarks has then symmetrized in sets of three and otherwise antisymmetrized. This can be described in various equivalent ways. One is to consider "para-Fermi statistics of rank 3"[6] and then to impose the restriction that all physical particles be fermions or bosons; the quarks are then fictitious (i.e., always bound) and all physical three-quark systems are totally symmetric overall. An equivalent description, easier to follow, involves introducing nine types of quarks, that is, the usual three types in each of three "colors," say red, white, and blue. The restriction is then imposed that all physical states and all observable quantities like the currents be singlets with respect to the SU(3) of color (i.e., the symmetry that manipulates the color index). Again, the quarks are fictitious. Let us refer to this type of statistics as "quark statistics."

If we take the quark statistics seriously and apply it to current quarks as well as constituent quarks, then the closed-loop processes in the models are multiplied by a factor of 3, and the asymptotic ratio $\sigma(e^+e^-\rightarrow\text{hadrons})/\sigma(e^+e^-\rightarrow\mu^+\mu^-)$ becomes $3\cdot\frac{2}{3}=2$.

Experiments at present are too low in energy and not accurate enough to test this prediction, but in the next year or two the situation should change. Meanwhile, is there any supporting evidence? Assuming that the connected light-cone algebra is right, we should like to know whether we can abstract the disconnected part as well, and whether the statistics are right. In fact, there is evidence from the decay of the π^0 into 2γ. It is well known that in the partially conserved axial current (PCAC) limit, with $m_\pi^2\rightarrow 0$, Adler and others[7] have given an exact formula for the decay amplitude $\pi^0\rightarrow 2\gamma$ in a "quark-gluon" model theory. The amplitude is a known constant times $(\sum Q_{1/2}^2-\sum Q_{-1/2}^2)$, where the sum is over the types of quarks and the charges $Q_{1/2}$ are those of $I_z=\frac{1}{2}$ quarks, while the charges $Q_{-1/2}$ are those of $I_z=-\frac{1}{2}$ quarks. The amplitude agrees with experiment, within the errors, in both sign and magnitude if $\sum Q_{1/2}^2-\sum Q_{-1/2}^2=1$.[8] If we had three Fermi-Dirac quarks, we would have $(\frac{2}{3})^2-(-\frac{1}{3})^2=\frac{1}{3}$, and the decay rate would be wrong by a factor of $\frac{1}{9}$. With "quark statistics," we get $\frac{1}{3}\cdot 3=1$ and everything is all right, assuming that PCAC is applicable.

There is, however, the problem of the derivation of the Adler formula. In the original derivation a renormalized perturbation expansion is applied to the "quark-gluon" model theory, and it is shown that only the lowest-order closed-loop diagram survives in the PCAC limit,[9] so that an exact expression can be given for the decay amplitude. Clearly this derivation does not directly suit our purposes, since our light-cone algebra is not obtainable by renormalized perturbation theory term by term. Of course, the situation might change if all orders are summed.

Recently it has become clear that the formula can be derived without direct reference to renormalized perturbation theory, from considerations of light-cone current algebra. Crewther has contributed greatly to clarifying this point,[10] using earlier work of Wilson[11] and Schreier.[12] Our objectives in this chapter are to call attention to Crewther's work, to sketch a derivation that is somewhat simpler than his, and to clarify the question of statistics.

We assume the connected light-cone algebra, and we make the further abstraction, from free quark theory or formal "quark-gluon" theory, of the principle that not only commutators but also products and physically ordered products of current operators obey scale invariance near the light cone, so that, apart from possible subtraction terms involving four-dimensional δ functions, current products near the light cone are given by the same formula as current commutators, with the singular functions changed from $\epsilon(z_0)\delta[(z^2)]$ to $(z^2 - i\epsilon z_0)^{-1}$ for ordinary products or $(z^2 - i\epsilon)^{-1}$ for ordered products.

Then it can be shown from consistency arguments that the only possible form for the disconnected parts (two-, three-, and four-point functions) is that given by free quark theory or formal "quark-gluon" theory, with only the coefficient needing to be determined by abstraction from a model. (In general, of course, the coefficient could be zero, thus changing the physics completely.) Then, from the light-cone behavior of current products, including connected and disconnected parts, the Adler formula for $\pi^0 \to 2\gamma$ in the PCAC limit can be derived in terms of that coefficient.

If we take the coefficient from the model with "quark statistics," predicting the asymptotic ratio of $\sigma(e^+e^- \to \text{hadrons})/\sigma(e^+e^- \to \mu^+\mu^-)$ to be 2 for one-photon annihilation, we obtain the correct value of the $\pi^0 \to 2\gamma$ decay amplitude, agreeing with experiment in magnitude and sign. Conversely, if for any reason we do not like to appeal to the model, we can take the coefficient from the observed $\pi^0 \to 2\gamma$ amplitude and predict in that way that the asymptotic value of $\sigma(e^+e^- \to \text{hadrons})/\sigma(e^+e^- \to \mu^+\mu^-)$ should be about 2.

Some more complicated and less attractive models that agree with the observed $\pi^0 \to 2\gamma$ amplitude are discussed in Section 3.

2. LIGHT-CONE ALGEBRA

The ideas of current algebra stem essentially from the attempt to abstract, from field theoretic quark models with interactions, certain algebraic relations obeyed by weak and electromagnetic currents to all orders in the strong interaction, and to postulate these relations for the system of real hadrons, while suggesting possible experimental tests of their validity. In four dimensions, with spinor fields involved, the only renormalizable models are ones that are barely renormalizable, such as a model of spinors coupled to a neutral vector "gluon" field. Until recently, the relations abstracted, such as the equal-time commutation relations of vector and axial charges or charge densities, were true in each order of renormalized perturbation theory in such a model. Now, however, one is considering the abstraction of results that are true only formally, with canonical manipulation of operators, and that fail, by powers of logarithmic factors, in each order of renormalized perturbation theory, in all barely renormalizable models (although they might be all right in a super-renormalizable model, if there were one).

The reason for the recent trend is, of course, the tendency of the deep inelastic electron scattering experiments at SLAC to encourage belief in Bjorken scaling, which fails to every order of renormalized perturbation theory in barely renormalizable models. There is also the availability of beautiful algebraic results, with Bjorken scaling as one of their predictions, if formal abstractions are accepted. The simplest such abstraction is that of the formula giving the leading singularity on the light cone of the connected part of the commutator of the vector or axial vector currents,[1] for example:

$$\begin{aligned}
[F_{i\mu}(x),F_{j\nu}(y)] &\hat{=} [F_{i\mu}{}^5(x),F_{j\nu}{}^5(y)] \\
&\hat{=} \frac{1}{4\pi}\partial_\rho\left\{\epsilon(x_0-y_0)\delta\left[(x-y)^2\right]\right\} \\
&\quad\times\left\{(if_{ijk}-d_{ijk})\left[s_{\mu\nu\rho\sigma}F_{k\sigma}(y,x)+i\epsilon_{\mu\nu\rho\sigma}F_{k\sigma}{}^5(y,x)\right]\right. \\
&\quad\left.+(if_{ijk}+d_{ijk})\left[s_{\mu\nu\rho\sigma}F_{k\sigma}(x,y)-i\epsilon_{\mu\nu\rho\sigma}F_{k\sigma}{}^5(x,y)\right]\right\} \qquad (1)
\end{aligned}$$

On the right-hand side we have the connected parts of bilocal operators $F_{i\mu}(x,y)$ and $F_{i\mu}{}^5(x,y)$, which reduce to the local currents $F_{i\mu}(x)$ and $F_{i\mu}{}^5(x)$ as $x\to y$. The bilocal operators are defined as observable quantities only in the vicinity of the light-cone, $(x-y)^2=0$. Here

$$s_{\mu\nu\rho\sigma}=\delta_{\mu\rho}\delta_{\nu\sigma}+\delta_{\nu\rho}\delta_{\mu\sigma}-\delta_{\mu\nu}\delta_{\rho\sigma}.$$

Formula 1 gives Bjorken scaling by virtue of the finite matrix elements assumed for $F_{i\mu}(x,y)$ and $F_{i\mu}{}^5(x,y)$; in fact, the Fourier transform of the matrix element of $F_{i\mu}(x,y)$ is just the Bjorken scaling function. The fact that all charged fields in the model have spin $\frac{1}{2}$ determines the algebraic structure of the formula and gives the prediction $(\sigma_L/\sigma_T)_{Bj}\to 0$ for deep inelastic electron scattering, not in contradiction with experiment. The electrical and weak charges of the quarks in the model determine the coefficients in the formula, and give rise to numerous sum rules and inequalities for the SLAC-MIT experiments in the Bjorken limit, again none in contradiction with experiment.

The formula for the leading light-cone singularity in the commutator contains, of course, the physical information that near the light cone we have full symmetry with respect to SU(3)×SU(3) and with respect to scale transformations in coordinate space. Thus there is conservation of dimension in the formula, with each current having $l=-3$ and the singular function $x-y$ also having $l=-3$.

A simple generalization of the abstraction that we have considered turns into a closed system, called the basic light-cone algebra. Here we commute the bilocal operators as well, for instance, $F_{i\mu}(x,u)$ with $F_{j\nu}(y,v)$, as all of the six intervals among the four space-time points approach 0, so that all four points tend to lie on a lightlike straight line in Minkowski space. Abstraction from the model gives us, on the right-hand side, a singular function of one coordinate difference, say $x-v$, times a bilocal current $F_{i\alpha}$ or $F_{i\alpha}{}^5$ at the other two points, say y and u, plus an expression with (x,v) and (y,u) interchanged, and the system closes algebraically. The formulas are just like Eq. 1. We shall assume here the validity of the basic light-cone algebraic system, and discuss the possible generalization to products and to disconnected parts. In Section 4, we conclude from the generalization to products that the form of an expression like $\langle \mathrm{vac}|F_{i\alpha}(x)\, F_{j\beta}(y,z)|\mathrm{vac}\rangle$ for disconnected parts is uniquely determined from the consistency of the connected light-cone algebra to be a number N times the corresponding expression for three free Fermi-Dirac quarks, when x,y, and z tend to lie on a straight lightlike line. The $\pi^0\to 2\gamma$ amplitude in the PCAC approximation is then calculated in terms of N and is proportional to it. Thus we do not want N to be zero.

The asymptotic ratio $\sigma(e^+e^-\to\text{hadrons})/\sigma(e^+e^-\to\mu^+\mu^-)$ from one-photon annihilation is also proportional to N. We may either determine N from the observed $\pi^0\to 2\gamma$ amplitude and then compute this asymptotic ratio approximately, or else appeal to a model and abstract the exact value of N, from which we calculate the amplitude of $\pi^0\to 2\gamma$. In a model, N depends on the statistics of the quarks, which we discuss in the next section.

3. STATISTICS AND ALTERNATIVE SCHEMES

As we remarked in Section 1, the presumably unwanted Fermi-Dirac statistics for the quarks, with $N=1$, would give $\sigma(e^+e^-\to\text{hadrons})/\sigma(e^+e^-\to\mu^+\mu^-)\to 2/3$. (Such quarks could be real particles, if necessary.) Now let us consider the case of "quark statistics," equivalent to para-Fermi statistics of rank 3 with the restriction that all physical particles be bosons or fermions. (Quarks are then fictitious, permanently bound. Even if we applied the restriction only to baryons and mesons, quarks would still be fictitious, as we can see by applying the principle of cluster decomposition of the S-matrix.)

The quark field theory model or the "quark-gluon" model is set up with three fields, q_R, q_B, and q_W, each with three ordinary SU(3) components, making nine in all. Without loss of generality, they may be taken to anticommute with one another as well as with themselves. The currents all have the form $\bar{q}_R q_R + \bar{q}_B q_B + \bar{q}_W q_W$, and are singlets with respect to the SU(3) of color. The physical states too are restricted to be singlets under the color SU(3). For example, the $q\bar{q}$ configuration for mesons is only $\bar{q}_R q_R + \bar{q}_B q_B + \bar{q}_W q_W$, and the qqq configuration for baryons is only $q_R q_B q_W - q_B q_R q_W + q_W q_R q_B - q_R q_W q_B + q_B q_W q_R - q_W q_B q_R$. Likewise all the higher configurations for baryons and mesons are required to be color singlets.

We do not know how to incorporate such restrictions on physical states into the formalism of the "quark-gluon" field theory model. We assume without proof that the asymptotic light-cone results for current commutators and multiple commutators are not altered. Since the currents are all color singlets, there is no obvious contradiction.

The use of quark statistics then gives $N=3$ and $\sigma(e^+e^-\to\text{hadrons})/\sigma(e^+e^-\to\mu^+\mu^-)\to 2$. This is the value that we predict.

We should, however, examine other possible schemes. First, we might treat actual para-Fermi statistics of rank 3 for the quarks without any further restriction on the physical states. In that case, there are excited baryons that are not fermions and are not totally symmetric in the $3q$ configuration; there are also excited mesons that are not bosons. Whether the quarks can be real in this case without violating the principle of "cluster decomposition" (factorizing of the S-matrix when a physical system is split into very distant subsystems) is a matter of controversy; probably they cannot. In this situation, N is presumably still 3.

Another situation with $N=3$ is that of a physical color SU(3) that can really be excited by the strong interaction. Excited baryons now exist that are in octets, decimets, and so on with respect to color, and mesons in octets and higher configurations. Many conserved quantum numbers exist, and new interactions may have to be introduced to violate them. This is a wildly speculative scheme. Here the nine quarks can be real if necessary, that is, capable of being produced singly or doubly at finite energies and identified in the laboratory.

We may consider a still more complicated situation in which the relationship of the physical currents to the current nonet in the connected algebra is somewhat modified, namely, the Han-Nambu scheme.[13] Here there are nine quarks, capable of being real, but they do not have the regular quark charges. Instead, the u quarks have charges $1,1,0$, averaging to $\frac{2}{3}$; the d quarks have charges $0,0,-1$, averaging to $-\frac{1}{3}$; and the s quarks also have charges $0,0,-1$, averaging to $-\frac{1}{3}$. In this scheme, not only can the analog of the color variable really be excited, but also it is excited even by the electromagnetic current, which is no longer a "color" singlet. Since the expressions for the electromagnetic current in terms of the current operators in the connected algebra are modified, this situation cannot be described by a value of N. It is clear, however, from the quark charges, that the asymptotic behavior of the disconnected part gives, in the Han-Nambu scheme, $\sigma(e^+e^-\to\text{hadrons})/\sigma(e^+e^-\to\mu^+\mu^-)\to 4$. Because the formulas for the physical currents are changed, numerical predictions for deep inelastic scattering are altered too. For example, instead of the inequality $\frac{1}{4}\leqslant[F^{en}(\xi)/F^{ep}(\xi)]\leqslant 4$ for deep inelastic scattering of electrons from neutrons and protons, we would have $\frac{1}{2}\leqslant[F^{en}(\xi)/F^{ep}(\xi)]\leqslant 2$. However, comparison of asymptotic values with experiment in this case may not be realistic at the energies now being explored. The electromagnetic current is not a color singlet; it directly excites the new quantum numbers, and presumably the asymptotic formulas do not become applicable until above the thresholds for the new kinds of particles. Thus, unless and until entirely new phenomena are detected, the Han-Nambu scheme really has little predictive power.

A final case to be mentioned is one in which we have ordinary "quark statistics" but the usual group SU(3) is enlarged to SU(4) to accomodate a "charmed" quark u' with charge $\frac{2}{3}$ which has no isotopic spin or ordinary strangeness but does have a nonzero value of a new conserved quantum number, charm, which would be violated by weak interactions (in such a way as to remove the strangeness-changing part from the commutator of the hadronic weak charge operator with its Hermitian conjugate). Again the expression for the physical currents in terms of our connected algebra is altered, and again the asymptotic value of $\sigma(e^+e^-\to\text{hadrons})/\sigma(e^+e^-\to\mu^+\mu^-)$ is changed, this time to $[(\frac{2}{3})^2+(-\frac{1}{3})^2+(-\frac{1}{3})^2+(\frac{2}{3})^2]\cdot 3=\frac{10}{3}$. Just as in the Han-Nambu scheme, the predictive power is very low here until the energy is above the threshold for making "charmed" particles.

We pointed out in Section 1 that for three Fermi-Dirac quarks the Adler amplitude is too small by a factor of 3. For all the other schemes quoted above, however, it comes out just right and the decay amplitude of $\pi^0 \to 2\gamma$ in the PCAC limit agrees with experiment. One may verify that for all of these schemes $\sum Q_{1/2}{}^2 - \sum Q_{-1/2}{}^2 = 1$.The various schemes are summarized in the following table.

Scheme	$(e^+e^- \to \text{hadrons})$ / $(e^+e^- \to \mu^+\mu^-)$	Can quarks be real?
"Quark statistics"	2	No
Para-Fermi statistics rank 3	2	Probably not
Nine Fermi-Dirac quarks	2	Yes
Han-Nambu, Fermi-Dirac	4	Yes
Quark statistics + charm	10/3	No
Para-Fermi, rank 3 + charm	10/3	Probably not
Twelve Fermi-Dirac + charm	10/3	Yes

In what follows, we shall confine ourselves to the first scheme, as requiring the least change in the present experimental situation.

4. DERIVATION OF THE $\pi^0 \to 2\gamma$ AMPLITUDE IN THE PCAC APPROXIMATION

In the derivation sketched here, we follow the general idea of Wilson's and Crewther's method. We lean more heavily on the connected light-cone current algebra, however, and we do not need to assume full conformal invariance of matrix elements for small values of the coordinate differences.

To discuss the $\pi^0 \to 2\gamma$ decay in the PCAC approximation, we shall need an expression for

$$\langle \text{vac} | F_{e\alpha}(x) F_{e\beta}(y) F_{3\gamma}{}^5(x) | \text{vac} \rangle$$

when $x \approx y \approx z$. (Here e is the direction in SU(3) space of the electric charge.) In fact, we shall consider general products of the form

$$\langle \mathrm{vac}|F(x_1)F(x_2)\cdots F(x_n)|\mathrm{vac}\rangle$$

where F's stand for components of any of our currents, and we shall examine the leading singularity when $x_1, x_2, \ldots, x_n$ tend to lie among a single lightlike line. (The case when they tend to coincide is then a specialization.)

We assume not only the validity of the connected light-cone algebra, which implies scale invariance for commutators near the light cone, but also scale invariance for products near the lightcone, with leading dimension $l = -3$ for all currents. There may be subtraction terms in the products, or at least in physical ordered products, for example, subtractions corresponding to four-dimensional δ functions in coordinate space; these are often determined by current consrvation. But apart from the subtraction terms the current products near the light cone have no choice, because of causality and their consequent analytic properties in coordinate space, but to obey the same formulas as the commutators, with $i\pi\epsilon(z_0)\delta(z^2)$ replaced by $\frac{1}{2}(z^2 - iz_0\epsilon)^{-1}$ for products and $\frac{1}{2}(z^2 - i\epsilon)^{-1}$ for physical ordered products.

Our general quantity $\langle \mathrm{vac}|F(x_1)F(x_2)\cdots F(x_n)|\mathrm{vac}\rangle$ may now be reduced, using successive applications of the product formulas near the light cone and ignoring possible subtraction terms, since all the intervals $(x_i - x_j)^2$ tend to zero, as they do when all the points x_i tend to lie on the same lightlike line.

A contraction between two currents $F(x_i), F(x_j)$ gives a singular function $S(x_i - x_j)$ times a bilocal $F(x_i x_j)$. If we now contract another local current with the bilocal, we obtain $S(x_i - x_j)S(x_k - x_j)F(x_i, x_k)$ and so on.

As long as we do not exhaust the currents, our intermediate states have particles in them and we are using the connected algebra generalized to products. Finally, we reach the stage where we have a string of singular functions multiplied by $\langle \mathrm{vac}|F(x_i, x_j)F(x_k)|\mathrm{vac}\rangle$, and the last contraction amounts to knowing the disconnected matrix element of a current product. However, the leading singularity structure of this matrix element ca. also be determined from the light-cone algebra by requiring consistent reductions of the three current amplitudes.

We can algebraically reduce a three-current amplitude in two possible ways. For each reduction, the algebra implies the existence of a known light-cone singularity. The reductions may also be carried out for an amplitude with a different ordering of the currents. One reduction of this amplitude yields the same two-point function as before, whereas the other reduction implies the existence of a second singularity in the two-point function. Hence we may conclude that the leading singularity of the two-point function when all points tend to a light line is given by the product of the two singularities identified by these reductions. Similarly, the leading singularity of the three-current amplitude is given by the product of the three singularities indicated by the different reductions. Since the connected light-cone algebra can be abstracted from the free quark model, the result of this analysis implies that the leading singularities of the two- and three-point functions are also given by the free quark model (say, with Fermi-Dirac quarks) and the only undetermined parameter is an overall factor, N, by which all vacuum amplitudes must be multiplied.

Since the singularity structure of the two-point function is determined, we can identify at least a part of the leading light line singularity of the n current amplitudes. Each different reduction of the n current amplitudes implies free quark singularities associated with this reduction. For two, three, and four current amplitudes, all of the singularities can be directly determined from the different reductions. For the five and higher-point functions not all of the singularities can be directly determined, but it is plausible that these others also have the free quark structure.

For the asymptotic value of $\sigma(e^+e^- \to \text{hadrons})/\sigma(e^+e^- \to \mu^+\mu^-)$, we are interested in the vacuum expected value of the commutator of two electromagnetic currents, and it comes out equal to N times a known quantity. Similarly, more complicated experiments testing products of four currents, for example, e^+e^- annihilation into hadrons and a massive muon pair or "γ"$-$"γ" annihilation into hadrons, might be considered. Also these processes are, in the corresponding deep inelastic limit, completely determined by the number N.

Returning to $\pi^0 \to 2\gamma$ in the PCAC approximation, we have $\langle \text{vac}|F_{e\alpha}(x) F_{e\beta}(y) F_{3\gamma}{}^5(z)|\text{vac}\rangle$ as the three space-time points approach a lightlike line, apart from subtraction terms, in terms of N times a known quantity. We now need only appeal to Wilson's argument (as elaborated by Crewther). The vacuum expected value of the physically ordered product $T(F_{e\alpha}(x), F_{e\beta}(y), \partial_\gamma F_{3\gamma}{}^5(z))$, taken at low frequencies, is what we need for the $\pi^0 \to 2\gamma$ decay with PCAC, and the Wilson-Crewther argument shows that it is determined from the small-distance behavior of $\langle \text{vac}|F_{e\alpha}(x) F_{e\beta}(y) F_{3\gamma}{}^5(z) |\text{vac}\rangle$, with the subtraction terms (which are calculable from current conservation in this case) playing no rôle. This remarkable superconvergence result, that the low-frequency matrix element can be calculated from a surface integral around the leading short-distance singularity (which is the same as the singularity if all three points tend to a lightlike line), makes possible the derivation of $\pi^0 \to 2\gamma$ in the PCAC approximation from the light-cone current algebra. We come out with the Adler result (i.e., the result for three Fermi-Dirac quarks) multiplied by N.

Thus the connected light-cone algebra provides a link between the $\pi^0 \to 2\gamma$ decay and the asymptotic ratio $\sigma(e^+e^- \to \text{hadrons})/\sigma(e^+e^- \to \mu^+\mu^-)$. Of course, one might doubt the applicability of PCAC to π^0 decay, or to any process in which other currents are present in addition to the axial vector current connected to the pion by PCAC. If the connected algebra is right, including products, then failure of the asymptotic ratio of the e^+e^- cross sections to approach the value 2 would be attributed either to such a failure of PCAC when other currents are present or else to the need for an alternative model such as we discussed in Section 3.

As a final remark, let us mention the "finite theory approach," as discussed in ref. 4 in connection with the light-cone current algebra. Here the idea is to abstract results not from the formal "quark-gluon" field theory model, but rather from the sum of all orders of perturbation theory (insofar as that can be studied) under two special assumptions. The assumptions are that the equation for the renormalized coupling constant that allows for a finite coupling constant renormalization has a root and that the value of the renormalized coupling constant is that root. Under these conditions, the vacuum expected values of at least some current products are less singular than in the free theory. Since the Adler result still holds in the "finite theory case," the connected light-cone algebra would have to break down. In particular, the axial vector current appearing in the commutator of certain vector currents is multiplied by an infinite constant. There are at present two alternative possibilities for such a "finite theory":

1. Only vacuum expected values of products of singlet currents are less singular than in the free theory;[14] only the parts of the algebra that involve singlet currents are wrong (e.g., the bilocal singlet axial vector current is infinite); the e^+e^- annihilation cross section would still behave scale invariantly.

2. All vacuum expected values of current products are less singular than in the free theory; the number N is zero; all bilocal axial vector currents are infinite; the e^+e^- annhilation cross section would decrease more sharply at high energies than in the case of scale invariance.

1. ACKNOWLEDGMENTS

For discussions, we are indebted to D. Maison, B. Zumino, and other members of the staff of the Theoretical Study Division of CERN. We are pleased to acknowledge also the hospitality of the Theoretical Study Division.

REFERENCES

1. H. Fritzsch and M. Gell-Mann, "Proceedings of the Coral Gables Conference on Fundamental Interactions at High Energies, January 1971," in *Scale Invariance and the Light Cone*, Gordon and Breach, New York, (1971).
2. H.J. Lipkin and S. Meshkov, *Phys. Rev. Letters*, **14**, 670 (1965).
3. R. Dashen and M. Gell-Mann, *Phys. Letters*, **17**, 142 (1965).
4. H. Fritzsch and M. Gell-Mann, *Proceedings of the International Conference on Duality and Symmetry in Hadron Physics*, Weizmann Science Press of Israel, Jerusalem, 1971.
5. G. Zweig, CERN Preprints TH. 401 and 412 (1964).
6. See, for example, O. W. Greenberg, *Phys. Rev. Letters*, **13**, 598 (1964).
7. S. L. Adler, *Phys. Rev.*, **177**, 2426 (1969); J. S. Bell and R. Jackiw, *Nuovo Cimento*, **60A**, 47 (1969).
8. S. L. Adler, in *Lectures on Elementary Particles and Fields* (1970 Brandeis University Summer Institute), MIT Press, Cambridge, Mass., 1971, and references quoted therein.
9. S. L. Adler and W. A. Bardeen, *Phys. Rev.*, **182**, 1517 (1969).
10. R. J. Crewther, Cornell preprint (1972).
11. K. G. Wilson, *Phys. Rev.*, **179**, 1499 (1969).
12. E. J. Schreier, *Phys. Rev. D*, **3**, 982 (1971).
13. M. Han and Y. Nambu, *Phys. Rev.*, **139**, 1006 (1965).
14. See also B. Schroer, Chapter 3 in this volume (p. 42).

Current Algebra – Quarks and what else?

Harald Fritzsch

H. Fritzsch (✉)
Physik-Department, Ludwig-Maximilians-Universität Physik-Department, München, Germany
e-mail: fritzsch@mppmu.mpg.de

H. Fritzsch (ed.), *Murray Gell-Mann and the Physics of Quarks*, Classic Texts in the Sciences, https://doi.org/10.1007/978-3-319-92195-2_9

Current Algebra: Quarks and What Else?

Harald Fritzsch*†

and

Murray Gell–Mann†**

CERN, Geneva, Switzerland

Proceedings of the XVI International Conference on High Energy Physics, Chicago, 1972. Volume 2, p. 135 (J. D. Jackson, A. Roberts, eds.)

Abstract

After receiving many requests for reprints of this article, describing the original ideas on the quark gluon gauge theory, which we later named QCD, we decided to place the article in the e–Print archive.

*On leave from the Max–Planck–Institut für Physik und Astrophysik. München, Germany.

†Present address: Lauritsen Laboratory of High Energy Physics, California, Institut of Technology, Pasadena, California.

**John Simon Guggenheim Memorial Foundation Fellow.

I. Introduction

For more than a decade, we particle theorists have been squeezing predictions out of a mathematical field theory model of the hadrons that we don't fully believe – a model containing a triple of spin 1/2 fields coupled universally to a neutral spin 1 field, that of the "gluon". In recent years, the triplet is usually taken to be the quark triplet, and it is supposed that there is a transformation, presumably unitary, that effectively converts the current quarks of the relativistic model into the constituent quarks of the naive quark model of baryon and meson spectrum and couplings.

We abstract results that are true in the model to all orders of the gluon coupling and postulate that they are really true of the electromagnetic and weak currents of hadrons to all orders of the strong interaction. In this way we build up a system of algebraic relations, so–called current algebra, and this algebraic system gets larger and larger as we abstract more and more properties of the model.

In section III, we review briefly the various stages in the history of current algebra. The older abstractions are correct to each order of renormalized perturbation theory in the model[1], while the more recent ones, those of light cone current algebra, are true to all orders only formally[3]. We describe the results of current algebra[2] in terms of commutators on or near a null plane, say $x_3 + x_0 = 0$.

In section IV, we attempt to describe, in a little more detail, using null plane language, the system of commutation relations valid in a free quark model that are known to remain unchanged (at least formally) when the coupling to a vector "gluon" is turned on. These equations give us a formidable body of information about the hadrons and their currents, supposedly exact as far as the strong interaction is concerned, for comparison with experiment. However, they by no means exhaust the degrees of freedom present in the model; they do not yield an algebraic system large enough to contain a complete description of the hadrons. In an Appendix, the equations of Section IV are related to form factor algebra.

In Section V, we discuss how further commutation of the physical quantities arising from light cone algebra leads, in the model field theory, to results dependent on the coupling constant, to formulae in which gluon field strength operators occur in bilocal current operators proliferate. Only when these relations are included do we finally get an algebraic system that contains nearly all the degrees of freedom of the model. We may well ask, however, whether it is the right algebraic system. We discuss briefly how the complete description of the hadrons involves the specification and slight enlargement of this algebraic system, the choice of representation of the algebra that corresponds to the complete set of hadron states, and the form of the mass or the energy operator, which must be expressible in terms of the algebra when it is complete. The choice of representation may be dictated by the algebra, and if so that would justify the use of a quark and gluon Fock space by some "parton" theorists.

Finally, in Section VI, it is suggested that perhaps there are alternatives to the vector gluon model as sources of information or as clues for the construction of the true hadron theory. Assuming we have described the quark part of the model correctly, can we replace the gluons by something else? The "string" or "rubber band" formulation, in ordinary coordinate space, of the zeroth approximation to the dual resonance model, is suggested as an interesting example.

Before embarking on our discussion of current algebra, we discuss in Section II the crucial point that quarks are probably not real particles and probably obey special statistics, along with related matters concerning the gluons of the field theory model.

II. FICTITIOUS QUARKS AND "GLUONS" AND THEIR STATISTICS

We assume here that quarks do not have real counterparts that are detectable in isolation in the laboratory – they are supposed to be permanently bound inside the mesons and baryons. In particular, we assume that they obey the special quark statistics, equivalent to "para-Fermi statistics of rank three" plus the requirement that mesons always be bosons and baryons fermions. The simplest description of quark statistics involves starting with three triplets of quarks, called red, white, and blue, distinguished only by the parameter referred to as color. These nine mathematical entities all obey Fermi–Dirac statistics, but real particles are required to be singlets with respect to the SU_3 of color, that is to say combinations acting like

$$\bar{q}_R q_R + \bar{q}_B q_B + \bar{q}_W q_W \text{ or } q_R q_B q_W - q_B q_R q_W - q_R q_W q_B - q_W q_B q_R + q_W q_R q_B + q_B q_W q_R \,. \quad (1)$$

The assumption of quark statistics has been common for many years, although not necessarily described in quite this way, and it has always had the following advantage: The constituent quarks as well as current quarks would obey quark statistics, since the transformation between them would not affect statistics, and the constituent quark model would then assign the lowest-lying baryon states (56 representation) to a symmetrical spatial configuration, as befits a very simple model.

Nowadays there is a further advantage. Using the algebraic relations abstracted formally from the quark–gluon model, one obtains a formula for the π^0 decay amplitude in the PCAC approximation, one that works beautifully for quark statistics but would fail by a factor 3 for a single Fermi–Dirac triplet[4)].

We have the option, no matter how far we go in abstracting results from a field theory model, of treating only color singlet operators. All the currents, as well as the stress–energy–momentum tensor $\Theta_{\mu\nu}$ that couples to gravity and defines the theory, are color singlets. We may, if we like, go further and abstract operators with three quark fields, or four quark fields and an antiquark field, and so forth, in order to connect the vacuum with baryon states, but we still need select only those that are color singlets in order to connect all physical hadron states with one another.

It might be a convenience to abstract quark operators themselves, or other non–singlets with respect to color, along with fictitious sectors of Hilbert space with triality non–zero, but it is not a necessity. It may not even be much of a convenience, since we would then, in describing the spatial and temporal variation of these fields, be discussing a fictitious spectrum for each fictitious sector of Hilbert space, and we probably don't want to load ourselves with so much spurious information.

We might eventually abstract from the quark–vector–gluon field theory model enough algebraic information about the color singlet operators in the model to describe all the degrees of freedom that are present.

For the real world of baryons and mesons, there must be a similar algebraic system, which may differ in some respects from that of the model, but which is in principle knowable. The operator $\Theta_{\mu\nu}$ could then be expressed in terms of this system, and the complete Hilbert space of baryons and mesons would be a representation of it. We would have a complete theory of the hadrons and their currents, and we need never mention any operators other than color singlets.

Now the interesting question has been raised lately whether we should regard the gluons as well as the quarks as being non–singlets with respect to color[5)]. For example, they could form a color octet of neutral vector fields obeying the Yang–Mills equations. (We must, of course, consider whether it is practical to add a common mass term for the gluon in that case – such a mass term would show up physically as a term in $\Theta_{\mu\nu}$ other than the quark bare mass term. In the past, we have referred to such an additional term that violates scale invariance, but does not violate $SU_3 \times SU_3$ as δ and its dimension as l_δ. Nowadays, ways of detecting expected values of δ are emerging.)[6)].

If the gluons of the model are to be turned into color octets, then an annoying asymmetry between quarks and gluons is removed, namely that there is no physical channel with quark quantum numbers, while gluons communicate freely with the channel containing the ω and ϕ mesons. (In fact, this communication of an elementary gluon potential with the real current of baryon number makes it very difficult to believe that all the formal relations of light cone current algebra could be true even in a "finite" version of singlet neutral vector gluon field theory.)

If the gluons become a color octet, then we do not have to deal with a gluon field strength standing alone, only with its square, summed over the octet, and with quantities like $\bar{q}\,(\partial_\mu - ig o_A B_{A\mu})\, q$, where the σ's are the eight 3×3 color matrices for the quark and the B's are the eight gluon potentials.

Now, suppose we look at such a model field theory, with colored quarks and colored gluons, including the stress–energy–momentum tensor. Basically the questions we are asking are the following:

1. Up to what point does the algebraic system of the color singlet operators for the real hadrons resemble that in the model? What is it in fact?

2. Up to what point does the representation of the algebraic system by the Hilbert space of physical hadron states resemble that in the model? What is it in fact?

3. Up to what point does $\Theta_{\mu\nu}$, expressed in term of the algebraic system, resemble that in the model? What is it in fact?

The measure of our ignorance is that for all we know, the algebra of color singlet operators, the representation, and even the form of $\Theta_{\mu\nu}$ could be exactly as in the model! We don't yet know how to extract enough consequences of the model to have a decisive confrontation with experiment, nor can we solve the formal equations for large g.
If we were solving the equations of a model, the first question we would ask is: Are the quarks really kept inside or do they escape to infinity? By restricting physical states and interesting operators to color singlets only, we have to some extent begged that question. But it re-emerges in the following form:
With a given algebraic system for the color singlet operators, can we find a locally causal $\Theta_{\mu\nu}$ that yields a spectrum corresponding to mesons and baryons and antibaryons and combinations thereof, or do we find a spectrum (in the color singlet states) that looks like combinations of free quarks and antiquarks and gluons?
In the next three Sections we shall usually treat the vector gluon, for convenience, as a color singlet.

III. REVIEW OF CURRENT ALGEBRA

In this section we sketch the gradual extension of algebraic results abstracted from free quark theory that remain true, either in renormalized perturbation theory or else only formally, when the coupling to a neutral vector gluon field is turned on.
The earlier abstractions were of equal–time commutation relations of current components. It was soon found that useful sum rules could best be derived from these by taking matrix elements between hadron states of equal P_3 as $P_3 \to \infty$, selecting the "gluon" components of the currents (those with matrix elements finite in this limit rather than tending to zero), and adding the postulate that, in the sum over intermediate states in the commutator, only states of finite mass need be considered. Thus formulae like the Adler–Weisberger and Cabibbo–Radicati sum rules were obtained and roughly verified by experiment.
Nowadays, the same procedure is usually accomplished in a slightly different way that is a bit cleaner – the hadron momenta are left finite instead of being boosted by a limit of Lorentz transformations, and the equal time surface is transformed by a corresponding limit of Lorentz transformations into a null plane, with $x_3 + x_0 =$ constant, say zero. The hypothesis of saturation by finite mass intermediate states is replaced by the hypothesis that the commutation rules of good components can be abstracted from the model not only on an equal time plane, but on a null plane as well[7,8).
In the last few years, the process of abstraction has been extended to a large class of algebraic relations (those of "light cone current algebra") that are true only formally in the model, but fail to each order of renormalized perturbation theory - they would be true to each order if the model were super–renormalizable. The motivation has been supplied by the compatibility of the deep inelastic electron scattering experiments performed at SLAC with the scaling predictions of Bjorken, which is the most basic feature of "light

cone current algebra". The Bjorken scaling limit $q^2 \to \infty, 2p \cdot q \to \infty, \xi \equiv q^2/(-2p \cdot q)$ finite) corresponds in coordinate space to the singularity on the light cone $(x-y)^2 = 0$ of the current commutator $[j(x), j(y)]$, and the relations of light cone current algebra are obtained by abstracting the leading singularity on the light cone from the field theory model. The singular function of $x - y$ is multiplied by a bilocal current operator $\Theta(x, y)$ that reduces to a familiar local current as $x - y \to 0$. The Bjorken scaling functions $F(\xi)$ are Fourier transforms of the expected values of the bilocal operators. Numerous predictions emerge from the relations abstracted from the quark–gluon model for deep inelastic and neutrino cross–sections. For example, the spin 1/2 character for the quanta bearing the charge in the model is reflected in the prediction $\sigma_L/\sigma_T \to 0$, while the charges of the quarks are reflected in the inequalities $1/4sF^{\text{en}}(\xi)/F^{\text{ep}}(\xi) \leq 4$. So far there is no clear sign of my contradiction between the formulae and the experimental results.
We may go further and abstract from the model also the light–cone commutators of bilocal currents, in the limit in which all the intervals among the four points approach zero, that is to say, when all four points tend to lie on a light–like line. The same bilocal operators then recur as coefficients of the singularity, and the algebraic system closes.
The light cone results can be reformulated in terms of the null plane. We consider a commutator of local currents at two points x and y and allow the two points to approach the same null plane, say

$$x_+ \equiv x_3 + x_0 = 0, y_+ \equiv y_3 + y_0 = 0 \tag{2}$$

As mentioned above, when both current components are "good", we obtain a local commutation relation on the null plane, yielding another good component, or else zero. But when neither component is good, there is a singularity of the form

$$\delta(x_+ - y_+) \tag{3}$$

and the coefficient is a bilocal current on the null plane. It is this singularity, arising from the light–cone singularity, that gives the Bjorken scaling.
On the null plane, with $x_+ = 0$, the three coordinates are the transverse spacelike coordinates x_1 and x_2 (called $x_\perp$) and the lightlike coordinate $x_- \equiv x_3 - x_0$. Our bilocal currents $O(u, y)$ on the nullplane are functions of four coordinates: x_-, y_- and $x_\perp = y_\perp$, since the interval between x and y is lightlike.
We may now consider the commutator of two bilocal currents defined on neighboring null planes (in each case with a lightlike interval between the two arguments of the bilocal current). Again, when neither current component is good, there is a δ–function singularity of the spacing between the two null planes and the coefficient is a bilocal current defined on the common limiting null plane. In this language, as before in the light cone language, the system of bilocal currents closes.
We may commute two good components of bilocal currents on the same null plane, and,

as for local currents, we obtain a good component on the right–hand side, without any δ–function singularity at coincidence of the two null planes. Thus the good components of the bilocal currents $O(u, y)$ form a Lie algebra on the null plane, a generalization of the old Lie algebra of local good components on the null plane (recovered by putting $x_- = y_-$).
Now, how far can we generalize this new Lie algebra on the null plane and still obtain exact formulae, formally true to all orders in the coupling constant, but independent of it, so that free quark formulae apply?
In the next section, we take up that question, but first we summarize the situation of current algebra on and near the null plane.

IV. SUMMARY OF LIGHT CONE AND NULL PLANE RESULTS

Let us now be a little more explicit. We are dealing with 144 bilocal quantities $\mathcal{F}_{j\alpha}, \mathcal{F}_{j\alpha}, S_j, F$ and $T_{j\alpha\beta}$ all functions of $x - y$ with $(x-y)^2 \to 0$. Let us select the 3–direction for our null planes. Then in the model we can set $B_+ \equiv B_3 + B_0 = 0$ for the gluon potential by a choice of gauge. The gauge–invariance factor exp ig $\int_y^x B \cdot dl$ for a straight line path on a null plane is just exp $\left[i\frac{g}{2}B_+(x_- - y_-)\right] = 1$. Thus we have simple correspondences between our quantities and operators in the model:

$$\mathcal{F}_{j\alpha}(x, y) \sim \frac{i}{2}\bar{q}(x)\lambda_j\gamma_\alpha q(y), \quad \text{etc.}$$

and we have introduced the notation $\mathcal{D}\left(x, y, \frac{i}{2}\lambda_j\gamma_\alpha\right)$, etc., where

$$\mathcal{D}(x, y, G) \sim \bar{q}(x)Gq(y) \sim q^+(x)(\beta G)q(y)\,. \tag{4}$$

We are dealing with $\mathcal{D}(x, y, G)$ for every (12×12) matrix G, with

$$\mathcal{F}^5_{j\alpha}(x, y) = \mathcal{D}\left(x, y, \frac{i}{2}\lambda_j\gamma_\alpha, \gamma_5\right) S_j(x, y) = \mathcal{D}\left(x, y, \frac{1}{2}\lambda_j\right), \tag{5}$$

$$P_j(x, y) = \mathcal{D}\left(x, y, \frac{i}{2}\lambda_j\gamma_5\right), \text{ and } \mathcal{T}_{j\alpha\beta}(x, y) = \mathcal{D}\left(x, y, \frac{i}{2}\lambda_j\sigma_{\alpha\beta}\right). \tag{6}$$

The good components, in the old equal–time $P_3 \to \infty$ language, were those with finite matrix elements between states of finite mass and $P_3 \to \infty$. By contrast, bad components were those with matrix elements going like $P_3{}^{-1}$ and terrible components those with matrix elements going like $P_3{}^{-2}$.

In the null plane language, good components are those for which βG is proportional to $1 + \alpha_3$; thus the 36 good components are $\mathcal{F}_{j+}, \mathcal{F}^5_{j+}, \mathcal{T}_{j1+}, \mathcal{T}_{j2+}$ for $j = 0 \ldots 8$. The terrible components are those for which βG is proportional to $1 - \alpha_3$, hence $\mathcal{F}_{j-}, \mathcal{F}^5_{j-}, \mathcal{T}_{j1}$, and $\mathcal{T}_{j2-}$. The rest are bad; they have βG anticommuting with α_3 so that α_3 is -1 on the left and +1 on the right or vice versa.
Now the leading light cone singularity in the commutator of two bilocals is just given by the formula

$$[(\mathcal{D}(x,y,G), \mathcal{D}(u,v,G'))] \hat{=} \mathcal{D}(x,v,iG\gamma_\mu G')\, \partial_\mu \Delta(y-u) - \mathcal{D}(u,y,iG'\gamma_\mu G)\, \partial_\mu \Delta(v-x), \tag{7}$$

with $\Delta(z) = (2\pi)^{-1} \varepsilon(z_0)\, \delta(z^2)$.

When we commute two operators with coordinates lying on neighboring null planes with separation Δx_+, a singularity of the type $\delta(\Delta x_+)$ appears (as we have mentioned in Section III) multiplied by a bilocal operator, with coordinates lying in the common null plane as $\Delta x_+ \to 0$, and it is this term that gives rise to Bjorken scaling. The term in question comes from the component $\frac{\partial}{\partial z_+}\Delta(z)$ in $\partial_\mu \Delta(z)$, and is thus multiplied by $\mathcal{D}(x,\nu,iG\gamma_+ G')$ and $\mathcal{D}(u,y,iG'\gamma_+ G)$. Now $\beta\,(iG\gamma_+ G') = (\beta G)(1-\alpha_3)(\beta G')$, so it is clear that the singular Bjorken scaling term vanishes for good–good and good–bad commutators. In the case of the other components, we have, schematically, [bad, bad] $\to$ good, [bad, terrible] $\to$ bad, and [terrible, terrible] $\to$ terrible for the Bjorken singularity.
The vector and axial vector local currents $\mathcal{F}_{j\alpha}(x,x)$ and $\mathcal{F}^5_{j\alpha}(x,x)$ occur, of course, in the electromagnetic and weak interactions. The local scalar and pseudoscalar currents occur in the divergences of the non–conserved vector and the axial vector currents, with coefficients that are linear combinations of the bare quark masses, m_u, m_d and m_s, treated as a diagonal matrix. (Here m_u would equal m_d if isotopic spin conservation were perfect, while the departure of m_s from the common value of m_u and m_d is what gives rise to SU_3 splitting; the non–vanishing of m is what breaks $SU_3 \times SU_3$).
We see that all the 144 bilocals are physically interesting, including the tensor currents, because they all occur in the commutators of these local V, A, S, and P densities as coefficients of the $\delta(\Delta x_+)$ singularity. Commuting a local scalar with itself or a local pseudoscalar with itself leads to the same bilocal as commuting a transverse component of a vector with itself, and thus the light cone commutator of current divergences is predicted to lead to Bjorken scaling functions that are proportional to those observed in the light cone commutation of currents, while the coefficients permit the experimental determination of the squares of the quark bare masses. Unfortunately, the relevant experiments are difficult. (The finiteness of the bare masses, as compared with the divergences encountered term in renormalized perturbation theory in a gluon model, presumably has the same origin as the scaling, which also fails term by term in renormalized perturbation theory.)

As we have outlined in Section III, we begin the construction of the algebraic system on the null plane by commuting the good bilocals with one another. The leading singularity on the light cone (Eq.(4.1)) gives rise to the simple closed algebra we have mentioned, but we need also the additional assumption that lower singularities on the light cone give no contribution to the good–good commutators on the null plane. This additional assumption can be squeezed out of the model in various ways. The simplest, however, is to use canonical quantization of the quark–gluon model on the null plane.
In the model, the quark field q is written as $q_+ + q_-$, where $q_\pm = \frac{1}{2}(1 \pm \alpha_3)\, q$. Then q_+ obeys the canonical rules $\{q_{+\alpha}(x), q_{+\beta}(y)\} = 0, \left\{q_{+\alpha}(x).q^+_{+\beta}(y)\right\} = \delta^{(3)}(x-y)\frac{1}{2}(l + a_3)_{\alpha\beta}$ on the null plane, where $\delta^{(3)}(x-y) = \delta(x_\perp - y_\perp)\,\delta(x_- - y_-)$. Thus for any good matrices βA_{++} and (βB_{++}), we have on the null plane

$$[\mathcal{D}(x, y, \beta A_{++}), \mathcal{D}(u, v, \beta B_{++})] =$$

$$\mathcal{D}(x, v\beta A_{++}B_{++})\,\delta^{(3)}(y-u) - \mathcal{D}(u, y, \beta B_{++}A_{++})\,\delta^{(3)}(v-x),$$

which is just what we would get from (4.1) with no additional contribution from lower light cone singularities.
The good–good commutation relations (4.2) on the null plane, together with the equations (4.1) for the leading light cone singularity in the commutator of two bilocal currents, illustrate how far we can go with abstracting free quark formulae that are formally unchanged in the model when the gluon coupling is turned on.
One may go further in certain directions. For example, the formulae for the leading light cone singularity presumably apply to disconnected as well as connected parts of matrix elements, and thus the question of the vacuum expected value of a bilocal operator arises. In the model, the coefficient of the leading singularity as $(x-y)^2 \to 0$ of such an expected value is formally independent of the coupling constant, and we abstract that as well – the answer here is dependent on statistics, however, and we assume the validity of quark statistics. Thus we obtain predictions like the following:

$$\sigma\left(e^+ + e^- \to \text{ hadrons}\right)/\sigma\left(e^+ + e^- \to \mu^+ + \mu^-\right) \to 2 \tag{8}$$

at high energy to lowest order in the fine structure constant.

The leading light cone singularity of an operator product, or of a physical order (T^*) product, may also be abstracted from the model, except for certain subtraction terms (often calculable and / or unimportant) that behave like four–dimensional δ-functions in coordinate space. To go from a commutator formula to a physical ordered product formula, we simply perform the substitutions

$$(2\pi)^{-1}\varepsilon(z)\delta\left(z^2\right) \to \left(4\pi^2 i\right)^{-1}\left(z^2 - iz_0\varepsilon\right)^{-1} \to \left(4\pi^2 i\right)^{-1}\left(z^2 - i\varepsilon\right)^{-1}. \tag{9}$$

With the aid of the product formulae and the vacuum expected values, we obtain the PCAC value of the $\pi^0 \to 2\gamma$ decay amplitude.

Other exact abstractions from the vector gluon model that do not contain g are divergence and curl relations for local V and A currents:

$$\begin{aligned} \frac{\partial}{\partial x_\mu}\mathcal{D}\left(x, x, \frac{i}{2}\lambda_i\gamma_\mu\right) &= \mathcal{D}\left(x, x, \frac{i}{2}[m, \lambda_i]\right), \\ \frac{\partial}{\partial x_\mu}\mathcal{D}\left(x, x, \frac{i}{2}\lambda_i\gamma_\mu\gamma_5\right) &= \mathcal{D}\left(x, x, \frac{i}{2}\{m, \lambda_i\}\gamma_5\right), \end{aligned} \tag{10}$$

but we also have, as presented elsewhere[2)],

$$\begin{aligned} \frac{\partial}{\partial x_\nu}\mathcal{D}\left(x, x, \frac{1}{2}\lambda_i\sigma_{\mu\nu}\right) &= -\mathcal{D}\left(x, x, \frac{i}{2}\{m, \lambda_i\}\gamma_\nu\right) \\ &\quad + \left[\left(\frac{\partial}{\partial x_\nu} - \frac{\partial}{\partial y_\nu}\right)\mathcal{D}\left(x, y, \frac{i}{2}\lambda_i\right)\right]_{x=y} \end{aligned} \tag{11}$$

$$\begin{aligned} \frac{\partial}{\partial x_\nu}\mathcal{D}\left(x, x, \frac{1}{2}\lambda_i\sigma_{\mu\nu}\gamma_5\right) &= -\mathcal{D}\left(x, x, \frac{i}{2}\right)[m, \lambda_i]\gamma_\nu\gamma_5 \\ &\quad + \left[\left(\frac{\partial}{\partial x_\nu} - \frac{\partial}{\partial y_\nu}\right)\mathcal{D}\left(x, y, \frac{i}{2}\lambda_i\gamma_5\right)\right] \end{aligned} \tag{12}$$

and a number of other formulae, including the following:

$$\left[\left(\frac{\partial}{\partial x_\nu} - \frac{\partial}{\partial y_\nu}\right)\mathcal{D}\left(x, y, \frac{i}{2}\lambda_i\gamma_\nu\right)\right]_{x=y} = \mathcal{D}\left(x, x, \frac{i}{2}\{\lambda_i, m\}\right) \tag{13}$$

In the last three formulae, it must be pointed out that for a general direction of $x - y$ we have the gauge–invariant correspondence

$$\mathcal{D}(x, y, G) \sim \bar{q}(x)Gq(y) \exp \mathrm{ig} \int_y^x B \cdot dl, \tag{14}$$

which is independent of the path from y to x when the coordinate difference and the path are taken as first order infinitesimals. The first internal derivative

$$\left[\left(\frac{\partial}{\partial x_\mu} - \frac{\partial}{\partial y_\mu}\right)\mathcal{D}(x, y, G)\right]_{x=y} \tag{15}$$

is physically interesting for all directions μ (and not just the – direction), as a result of Lorentz covariance.
In Eqs. (4.5–4.7), we have for the moment thrown off the restriction to a single null plane. In the next Section, we return to the consideration of the algebra on the null plane, and we see how further extensions give a much wider algebra, in which departures from free quark relations begin to appear.

V. THE FURTHER EXTENSION OF NULL PLANE ALGEBRA

We now look beyond the commutation relations of good bilocals on the null plane. In the model, then, we have to examine operators containing q_- or q_-^+ or both. The Dirac equation in the gauge we are using ($B_+ = 0$ on the null plane) tells us that we have

$$-2i\frac{\partial q_-}{\partial x_-} = (\alpha_\perp \cdot (-i\nabla_\perp - gB_\perp) + \beta m)\, q_+. \tag{16}$$

In terms of Eq. (5.1), we can review the various anticommutators on the null plane. We have already discussed the trivial one,

$$\left(q_+(x), q_+^\dagger(y)\right) = \delta(x_- - y_-) \cdot \frac{1}{2}(1+\alpha_3)\,\delta(x_\perp - y_\perp)\,. \tag{17}$$

Using (5.1), (5.2), the fact that $B_\perp$ commutes with q_+ on the null plane, and the equal–time anticommutator $\left\{q_-, q_+^\dagger\right\} = 0$, we obtain well–known result

$$\left\{q_-(x), q_+^\dagger(y)\right\} = \frac{i}{4}\varepsilon(x_- - y_-)\left[\alpha_\perp \cdot \left(i\nabla_\perp^{(y)} - gB_\perp(y)\right) + \beta m\right]\frac{1}{2}(1+\alpha_3)\,\delta(x_\perp - y_\perp)\,. \tag{18}$$

Using the same method a second time, one finds, for $y_- > x_-$,

$$\begin{aligned}\left\{q_-(x), q_-^\dagger(y)\right\} = &-\frac{1}{8}\int_{x_-}^{y_-} dr_-\left[\alpha_\perp\left(-i\nabla_\perp^{(x)} - gB_\perp(x_\perp, r_-)\right) + \beta m\right]^2\left(\frac{1-\alpha_3}{2}\right)\delta(x_\perp - y_\perp)\\ &+ i\frac{g^2}{32}\int_{x_-}^{y_-} dy'_-\int_{x_-}^{y'_-} dx'_-\left[\alpha_\perp q_+\left(x_\perp, x'_-\right); q_+\left(y_\perp, y'_-\right)\alpha_\perp\right]\delta(x_\perp - y_\perp)\\ &+ \delta(x_+ - y_+)\left(\frac{1-\alpha_3}{2}\right)\delta(x_\perp - y_\perp),\end{aligned} \tag{19}$$

where the singularity at the coincidence of the two null planes appears as an unpleasant integration constant. This singularity is, of course, responsible in the model for the Bjorken singularity in the commutator of two bad or terrible operators.
Because of the singularity, it is clumsy to construct the wider algebra by commuting all

our bilocals with one another. Instead, we adopt the following procedure. Whenever a bilocal operator corresponds to one in the model containing $q_-^+(x)$, we differentiate with respect to x_-; whenever it corresponds to one in the model containing $q_{(y)}$, we differentiate with respect to y_-. Thus we "promote" all our bilocals to good operators. We construct the wider algebra by starting with the original good bilocals and these promoted bad and terrible bilocals. We commute all of these, commute their commutators, and so forth, until the algebra closes. Then, later on, if we want to commute an unpromoted operator, we use the information contained in equations of the model like (5.1) - (5.3) to integrate over x_- or y_- or both and undo the promotion. (A similar situation obtains for operators corresponding to those in the model containing the longitudinal gluon potential B_-.)
Now let us classify the matrices βG into four categories:
the good ones, $\beta G = A_{++}$, with $\alpha_3 = 1$ on both sides;
the bad ones $\beta G = A_{+-}$ that have $\alpha_3 = 1$ on the left and -1 on the right;
the bad ones $\beta G = A_{-+}$ that have $\alpha_3 = -1$ on the left and $+1$ on the right;
and the terrible ones $\beta G = A_{--}$, with $\alpha_3 = -1$ on both sides.

Then, wherever q_- or q_-^+ appears, we promote the operator by differentiating q_- or q_-^+ with respect to its argument in the - direction. We obtain, then:

$$\mathcal{D}\left(x, y, \beta A_{++}\right),$$

the good operators, unchanged;

$\frac{\partial}{\partial x_-}\mathcal{D}\left(x, y, \beta A_{-+}\right)$ and $\frac{\partial}{\partial y_-}\left(x, y, \beta, A_{+-}\right)$ promoted bad operators:

and

$\frac{\partial}{\partial x_-}\frac{\partial}{\partial y_-}\mathcal{D}\left(x, y, \beta A_{--}\right)$, promoted terrible operators.

All 144 of these operators now are given, in the model, in terms of q_+ and q_+^+, but the promoted bad and terrible operators involve the expressions $\left(\nabla_\perp - igB_\perp\right)q_+$ and $\left(\nabla_\perp + igB_\perp\right)q_+^+$. In fact, substituting the Dirac equation for $\frac{\partial q_-}{\partial x_-}$ into the definitions of the promoted bad and terrible operators, we see that we obtain good operators (with coefficients depending on bare quark masses) and also good matrices sandwiched between $\left(\nabla_\perp + igB_\perp\right)q_+^+$ and q_+ or between q_+^+ and $\left(\nabla_\perp - igB_\perp\right)q_+$ or between $\left(\nabla_\perp + igB_\perp\right)q_+^+$ and $\left(\nabla_\perp - igB_\perp\right)q_+$.
The null plane commutators of all these operators with one another are finite, well-defined, and physically meaningful, but they lead to an enormous Lie algebra that is not identical with the one for free quarks, but instead contains nearly all the degrees of freedom of the model.
Let us first ignore any lack of commutation of the B's with one another. We keep commuting the operators in question with one another. When $\nabla_\perp \pm igB_\perp$ appears acting on a $\delta^{(3)}$

function, we can always perform an integration and fold it over onto an operator. Thus the number of applications of $\nabla_\perp \pm igB_\perp$ grows without limit. Since these gauge derivatives do not commute with one another, but give field strengths as commutators, it can easily be seen that we end up with all possible operators corresponding to $\bar{q}_+(x)Gq_+(y)$ acted on by any gauge invariant combination of transverse gradients and potentials. We have to put it differently, the operators corresponding to $\bar{q}_+(x)Gq_+(y)$ exp ig $\int_P B \cdot dl$ for any pair of points x and y on the null plane connected by any path P lying in the null plane. We could think of these as operators $\mathcal{D}(x, y, G, P)$ depending on the path P, with $\beta G = A_{++}$.

In fact the B's do not commute with another in the model, and so we get an even more complicated result. We have

$$[B_{\perp i}(x), B_{\perp j}(y)] \sim \varepsilon\,(x_- - y_-)\,\delta\,(x_\perp - y_\perp)\,\delta_{ij} \tag{20}$$

on the null plane, and the commutation of promoted bad and terrible bilocals with one another leads to operators corresponding to $\bar{q}_+(x)Gq_+(y)\bar{q}_+(a)G'q_+(b)$. Further commutation then introduces an unlimited number of sideways gradients, gluon field strengths, and additional quark pairs, until we end up with all possible operators of the model that can be constructed from equal numbers of $\bar{q}_+$'s and q_+'s at any points on the null plane and from exponentials of $ig \int B \cdot dl$ for any paths connecting these points.

If we keep track of color, we note that only color singlets are generated. If the gluons are a color octet Yang–Mills field, we must make suitable changes in the formalism but again we find that only color singlets are generated. The coupling constant g that occurs is, of course, the bare coupling constant. If may not be intrinsic to the algebraic system (equivalent to that of quarks and gluons) on the null plane, but it certainly enters importantly into the way we reach the system starting from well–known operators.

A troublesome feature of the extended null plane algebra is the apparent absence of operators corresponding to those in the model that contain only gluon field strengths and no quark operators; for a color singlet gluon, the field strength itself would be such an operator, while for a color octet gluon we could begin with bilinear forms in the field strength in order to obtain color singlet operators. Can we obtain these quark–free operators by investigating discontinuities at the coincidence of coordinates characterizing quark and antiquark fields in the model? At any rate, we certainly want these quarkfree operators included in the extended algebra.

Now when our algebra has been extended to include the analogs of all relevant operators of the model on the null plane that are color singlets and have baryon number $A = 0$, then the Hilbert space of all physical hadron states with $A = 0$ is an irreducible representation of the algebra.

If we wish, we might as well extend the algebra further by including the analogs of color singlet operators of the model (on the null plane) that would change the number of baryons. In that case, the entire Hilbert space of all hadron states is an irreducible representation of the complete algebra. From now on, let us suppose that we are always dealing with the complete color singlet algebra (whether the one abstracted from the quark–gluon model or some other) and with the complete Hilbert space, which is an irreducible representation of it.

The representation may be determined by the algebra and the uniqueness of the physical vacuum. We note that we are dealing with arbitrarily multilocal operators, functions of any number of points on the null plane. We can Fourier transform with respect to all these variables and obtain Fourier variables $(k_+, k_\perp)$ in place of the space coordinates. Since $B_+ = 0$, there is no formal obstacle to thinking of each k_+ as being like the contribution of the individual quark, antiquark or gluon to the total $P_+ = \sum k_+$. Now $P_+ = 0$ for the vacuum, and for any other state we can get $P_+ = 0$ only by taking $P_z \to -\infty$. The same kind of smoothness assumption that allows scaling can allow us to forget about matrix elements to such infinite momentum states. In that case, we have the unique vacuum state of hadrons as the only state of $P_+ = 0$, while all others have $P_+ > 0$. All Fourier components of multilocal operators for which $\sum k_+ < 0$ annihilate the physical vacuum. (Note in the null plane formalism we do not have to deal with a fictitious "free vacuum" as in the equal–time formalism.) The Fourier components of multilocal operator with $\sum k_+ > 0$ act on the vacuum to create physical states, and the orthogonality properties of these states and the matrix elements of our operators sandwiched between them are determined largely or wholly by the algebra. The details have to be studied further to see to what extent the representation is really determined. (The vacuum expected values contain one adjustable parameter in the case of free quarks, namely the number of colors.)

Once we have the representation of the complete color singlet algebra on the null plane, as well as the algebra itself, then the physical states of hadrons can all be written as linear combinations of the normalized basis states of the representation. These coefficients represent a normalized set of Fock space wave functions for each physical hadron state, with orthogonality relations for orthogonal physical states. Since the matrix elements of all null plane operators between basis states are known, the matrix elements between physical states of bilocal currents or other operators of interest are all calculable in terms of the Fock space wave functions[9].

This situation is evidently the one contemplated by "parton" theorists such as Feynman and Bjorken; they suppose that we know the complete algebra, that it comes out to be a quark–gluon algebra, and that the representation is the familiar one, so that there is a simple Fock space of quark, antiquark, and gluon coordinates. In the Fourier transform, negative values of each k_+ correspond to destruction and positive values to creation.

Now the listing of hadron states by quark and gluon momenta is a long way from listing by meson and baryon moments. However, as long as we stick to color singlets, there is not necessarily any obstacle to getting one from the other by taking linear combinations. The operator $M^2 = -P^2 - P_+P_-$ has to be such that its eigenvalues correspond to meson and baryon configurations, and not to a continuum of quarks, antiquarks and gluons.

The important physical questions are whether we have the correct complete algebra and

representation, and what the correct form of $\Theta_{\mu\nu}$ or P_μ or M^2 is, expressed in terms of that algebra.
In the quark–gluon model we have $\Theta_{\mu\nu} = \Theta^{\text{quark}}_{\mu\nu} + \Theta^{\text{glue}}_{\mu\nu}$, where

$$\begin{aligned}\Theta^{\text{quark}}_{\mu\nu} &= \frac{1}{4}\bar{q}\gamma_\mu\left(\partial_\nu - igB_\nu\right)q + \ldots q + \frac{1}{4}\bar{q}\gamma_\nu\left(\partial_\mu - igB_\mu\right)q \\ &-\frac{1}{4}\left(\partial_\mu + igB_\mu\right)\bar{q}\gamma_\nu q - \frac{1}{4}\left(\partial_\nu + igB_\nu\right)\bar{q}\gamma_\mu q\,, \end{aligned} \tag{21}$$

and $\Theta^{\text{glue}}_{\mu\nu}$ does not involve the quark variables at all. The term $\Theta^{\text{quark}}_{\mu\nu}$, by itself, has the wrong commutation rules to be a true $\Theta_{\mu\nu}$ (unless $g = 0$). For example, $\left(P^{\text{quark}}_1, P^{\text{quark}}_2\right) \neq 0$. The correct commutation rules are restored when we add the contribution from $\Theta^{\text{glue}}_{\mu\nu}$.
We can abstract from the quark–gluon model some or all the properties of $\Theta_{\mu\nu}$, in terms of the null plane algebra. We see that in the model we have

$$\Theta^{\text{quark}}_{++} = \left[\left(\frac{\partial}{\partial y_-} - \frac{\partial}{\partial x_-}\right)\mathcal{D}\left(x, y, \frac{1}{2}\gamma_+\right)\right]_{x=y} \tag{22}$$

and, as is well–known, the expected value of the right–hand side in the proton state can be measured by deep inelastic experiments with electrons and neutrinos. All indications are that it is not equal to the expected value of Θ_{++}, but rather around half of that, so that half is attributable to gluons, or whatever replaces them in the real theory.

In general, using the gauge–invariant definition of $\mathcal{D}$, we have in the model

$$\Theta^{\text{quark}}_{\mu\nu} = \left[\left(\frac{\partial}{\partial y_\nu} - \frac{\partial}{\partial x_\nu}\right)\mathcal{D}\left(x, y, \frac{1}{4}\gamma_\mu\right) + \left(\frac{\partial}{\partial y_\mu} - \frac{\partial}{\partial x_\mu}\right)\mathcal{D}\left(x, y, \frac{1}{4}\gamma_\nu\right)\right]_{x=y} \tag{23}$$

and Eq. (4.7) then gives us the obvious result

$$-\,\Theta^{\text{quark}}_{\mu\nu} = \mathcal{D}\left(x, x, m\right)\,. \tag{24}$$

Whereas in (5.5) we are dealing with an operator that belongs to the null plane algebra generated by good, promoted bad, and promoted terrible bilocal currents, other components of $\Theta^{\text{quark}}_{\mu\nu}$ are not directly contained in the algebra, neither are the bad and terrible local currents, nor their internal derivatives in directions other than $-$. In order to obtain the commutation properties of all these operators with those actually in the algebra, we must, as we mentioned above, undo the promotions by abstracting the sort of information contained in (5.3) and (5.4). Thus we are really dealing with a wider mathematical system than the closed Lie algebra abstracted from that of operators in the model containing $q^\dagger_+, q_+$ and $B_\perp$ only.

We shall assume that the true algebraic system of hadrons resembles that of the quark–gluon model at least to the following extent:

1) The null plane algebra of good components (4.2) and the leading light cone singularities (4.1) are unchanged.

2) The system acts as if the quarks had vectorial coupling in the sense that the divergence equation (4.3) and (4.4) are unchanged.

3) There is a gauge derivative of some kind, with path–dependent bilocals that for an infinitesimal interval become path–independent. Eqs. (4.5) - (4.7) are then defined and we assume they also are unchanged.

4) The expression (5.6) for $\Theta_{\mu\nu}^{\text{quark}}$ is also defined and we assume it, too, is unchanged, along with its corollary (5.7).

About the details of the form of the path–dependent null plane algebra arising from the successive application of gauge derivatives, we are much less confident, and correspondingly we are also less confident of the nature of the gluons, even assuming that we can decide whether to use a color singlet or a color octet. What we do assert is that there is some algebraic structure analogous to that in quark–gluon theory and that it is in principle knowable.

One fascinating problem, of course, is to understand the conditions under which we can have an algebra resembling that for quarks and gluons and yet escape having real quarks and gluons. Under what conditions do the bilocals act as if they were the products of local operators without, in fact, being seen. We seek answers to this and other questions by asking "Are there models other than the quark–gluon field theory from which we can abstract results? Can we replace $\Theta_{\mu\nu}^{\text{glue}}$ by something different and the gauge–derivative by a different gauge–derivative?"

VI. ARE THERE ALTERNATIVE MODELS?

In the search for alternatives to gluons, one case worth investigating is that of the simple dual resonance model. It can be considered in three stages: first, the theory of a huge infinity of free mesons of all spins; next, tree diagrams involving the interaction of these mesons; and finally loop diagrams. The theory is always treated as though referring to real mesons, and an S–matrix formulation is employed in which each meson is always on the mass shell.

Now the free stage of the model can easily be reformulated as a field theory in ordinary coordinate space, based on a field operator Φ that is a function not of one point in space, but of a whole path – it is infinitely multilocal. The free approximation to the dual resonance model is then essentially the quantum theory of a relativistic string or linear rubber band in ordinary space.

The coupling that leads, on the mass shell, to the tree diagrams of the dual resonance model has not so far been successfully reformulated as a field theory coupling but we shall assume that this can be done. Then the whole model theory, including the loops, would be a theory of a large infinity of local meson fields, all described simultaneously by a grand infinitely multilocal field Φ, couples to themselves and one another. The mesons, in the free approximation, lie on straight parallel Regge trajectories with a universal slope α'.

In the simplest form of such a theory, the grand field Φ (path) can be resolved into local fields $\phi(R), \Phi_{n\mu}(R), \Phi_{n\mu,n'\mu'}(R), \ldots$. There is a single scalar, a single infinity of vectors, a double infinity of tensors and scalars, and so forth. The matrices $a_{n\mu}$ and $a^+_{n\mu}$ of the dual theory connect these components of Φ with one another.

Perhaps the model theory of a gluon field can be replaced by a field theory version of a dual resonance model; the properties of operators, including $\Theta_{\mu\nu}$, would be abstracted from the new model instead of the old one. With $\alpha' \neq 0$, a term δ would naturally appear that violates scale invariance and is not related to the bare quark masses. (Probably $l_\delta = 0$ here rather than -2 as in the case of a gluon mass.) The gauge derivative in the other portion of $\Theta_{\mu\nu}$, referring to the quarks, would then involve a special linear combination of the $\Phi_{n\nu}(R)$ instead of the gluon potential $B_\mu(R)$.

An amusing point is that in the limit of a dual resonance theory as $\alpha' \to 0$ (so that the trajectories become flat), with attention concentrated on the value $\alpha = 1$, if the mathematics of a Lie group is built into the model, then the mass shell predictions become those of the corresponding massless Yang–Mills theory[10)]. That suggests that one might even try a dual resonance model as a replacement of a color octet Yang–Mills gluon model, with abstraction of the properties of color singlet operators.

We are not at all sure that what we are discussing here is a practical scheme, and if it is, we do not know how the resulting algebraic system differs from that of gluons. We put it forward merely in order to stimulate thinking about whether or not here are candidates for the algebra, the representation, and the form of $\Theta_{\mu\nu}$ other than those suggested by the gluon model.

Our attempt to use the dual model to construct a field theory has no bearing on whether the mass–shell dual model can lead to a complete S–matrix theory of hadrons; our suggestion resembles the use of limits of dual theories to obtain unified theories of weak and electromagnetic interactions or the theory of gravity.

One interesting speculation that is independent of what model we use for the stuff to which quarks are coupled is that perhaps when we perform the mathematical transformation from current quarks to constituent quarks and obtain the crude naive quark model of meson and baryon spectra and couplings, the gluons or whatever they are will also be approximately transformed into fictitious constituents, so that meson states would appear that act as if they were made of gluons rather than $q\bar{q}$ pairs. If there are indeed ten low–lying scalar mesons rather than nine, then we might interpret the tenth one (roughly speaking, the ε° meson) as the beginning of such a sequence of extra Su_3 singlet meson states. (A related question, much debated by specialists in the usual, mass–shell dual models, is whether the infinite sequence of meson and baryon Regge trajectories, all rising indefinitely and straight and parallel in zeroth approximation, should be extended to exotic channels, i. e., those with quantum numbers characteristic of $qqqq\bar{q}, q\bar{q}q\bar{q}$ etc.).

Let us end by emphasizing our main point, that it may well be possible to construct an explicit theory of hadrons, based on quarks and some kind of glue, treated as fictitious, but with enough physical properties abstracted and applied to real hadrons to constitute a complete theory. Since the entities we start with are fictitious, there is no need for any conflict with the bootstrap or conventional dual model point of view.

APPENDIX – BILOCAL FORM FACTOR ALGEBRA

We have described in Section III and IV a Lie algebra of good components of bilocal operators on a null plane. The generators are 36 functions of x_-, y_- and $x_\perp = y_\perp$, namely $\mathcal{F}_{j+}, \mathcal{F}^5_{j+}, \mathcal{T}_{jl+}$, and $\mathcal{T}_{j2+}$. We define $R \equiv 1/2(x+y)$ and $z \equiv x-y$; then we have functions of $R_\perp, R_-$, and z_-.
With z_- set equal to zero, we have just the usual good local operators on the null plane, related to the corresponding good local operators at equal times with $P_3 \to \infty$. We recall that in the early work using $P_3 \to \infty$ the most useful applications (fixed virtual mass sum rules) involved matrix elements with no change of longitudinal momentum, i. e., transverse Fourier components of the operators. Dashen and Gell-Mann[11)] studied these operators and found that between finite mass states their matrix elements do not depend separately on the transverse momenta of the initial and final states, but only on the difference, which is the Fourier variable $k_\perp$. Thus they obtained a "form factor algebra" generated by operators $F_i(k_\perp)$ and $F_i^5(k_\perp)$, to which, of course, one may adjoin $T_{il}(k_\perp)$ and $T_{i2}(k_\perp)$.
We may consider the analogous quantities using the null plane method and generating to bilocals:

$$F_i(k_\perp, z_-) \equiv$$

$$\int d^4R\delta(R_+)\mathcal{F}_{i+}(R, z_-)\ \exp\ ik_1\left[R_1 + P_+^{-1}(\Lambda_1 + J_2)\right]\ \exp\ ik_2\left[R_2 + P_+^{-1}(\Lambda_2 - J_1)\right] \tag{25}$$

and so forth. Here the integration over R_- assures us that $P_+ \equiv P_0 + P_3$ is conserved by the operator. (We note that Minkowski[12)] and others have studied the interesting problem of extracting useful sum rules from operators unintegrated over R_-, but we do not discuss that here.) The quantities $P_+^{-1}(\Lambda_1 + J_2)$ and $P_+^{-1}(\Lambda_2 - J_1)$ act like negatives of center-of-mass coordinates, $-\bar{R}_1$ and $-\bar{R}_2$, since on the null plane $x_+ = 0$ we have $\Lambda_1 + J_2 = -\int R_1\Theta_{++}d^4R\delta(R_+)$ and $\Lambda_1 + J_1 = -\int R_2\Theta_{++}d^4R\delta(R_+)$, while $P_+ = \int \Theta_{++}d^4R(R_+)$.
Our bilocal form factor algebra has the commutation rules

$$\left[F_i(k_\perp, z_-), F_j\left(k'_\perp, z'_-\right)\right] = i\, f_{ijk}F_k\left(k_\perp + k'_\perp, z_- + z'_-\right), \tag{26}$$

etc., where the structure constants in general are those of $[U_6]_w$. Putting $z_- = z_-{}' = 0$, we obtain exactly the form factor algebra of Dashen and Gell-Mann. If we specialize further to $k_\perp = k'_\perp = 0$, we obtain the algebra $[U_6]_{w,\infty,\ \text{currents}}$, of vector, axial vector, and tensor charges. It is not, of course, identical to the approximate symmetry algebra $[U_6]_{w,\infty\ \text{strong}}$, for baryon and meson spectra and vertices, but is related to it by a transformation, probably unitary. That is the transformation which we have described crudely as connecting current quarks and constituent quarks.
The behavior of the operators $F_i(k_\perp)$, etc., with respect to angular momentum in the

s–channel is complicated and spectrum–dependent; it was described by Dashen and Gell–Mann in their angular condition[10)]. There is a similar angular condition for the bilocal generalizations $F_i(k_\perp, z_-)$, etc.

The behavior of $F_i(k_\perp, z_-)$ and the other bilocals with respect to angular momentum in the cross–channel is, in contrast, extremely simple. If we expand $F_i(k_\perp, z_-)$ or $F_i^5(k_\perp, z_-)$ in powers of z_-, each power z_-^n corresponds to a single angular momentum, namely $J = n+1$.

As we expand $F_i(k_\perp, z_-)$, etc., in power series in z_-, we note that each term, in $z_-{}^{J-1}$, has a pole in $k_\perp^2$ at $k_\perp^2 + M^2 = 0$, where M is the mass of any meson of spin J. By an extension of the Regge procedure, we can keep $k_\perp^2$ fixed and let the angular momentum vary by looking at the asymptotic behavior of matrix elements of $F_i(k_\perp, z_-)$, etc., at large z_-. A Regge pole in the cross channel gives a contribution $z_-^{\alpha(-k_\perp^2)}\beta(k_\perp^2)$ $[\sin\ \pi\alpha(-k_\perp^2)]^{-1}$ and a cut gives a corresponding integral over α. Thus the bilocal form factor $F_i(k_\perp, z_-)$ couples to each Reggeon in the non–exotic meson system in the same way that $\mathcal{F}_i(k_\perp)$ couples to each vector meson. The contribution of each Regge pole to the asymptotic matrix element of $F_i(k_\perp, z_-)$ between hadron states A and B is given by the coupling of $\mathcal{F}_i(k_\perp, z_-)$ to that Reggeon multiplied by the strong coupling constant of the Reggeon to A and B.

It would be nice to substitute the Regge asymptotic behavior of $F_i(k_\perp, x_-)$ etc., into the commutation rules and obtain algebraic relations among the Regge residues. Unfortunately, the asymptotic limit is not approached uniformly in the different matrix elements, and the asymptotic Regge formulae cannot, therefore, be used for the operators everywhere in the equations (A.2); only partial results can be extracted.

References

1. M. Gell–Mann, Phys. Rev. 125, 1067 (1962) and Physics 1, 63 (1964).

2. H. Fritzsch, M. Gell–Mann, Proceedings of the Coral Gables Conference on Fundamental Interactions at High Energies, January 1971, in "Scale Invariance and the Light Cone", Gordon and Breach Ed. (1971), and Proceedings of the International Conference on Duality and Symmetry in Hadron Physics, Weizmann Science Press (1971).
J. M. Cornwall, R. Jackiw, Phys. Rev. D4, 367, (1971).
C.H. Llewellyn Smith, Phys. Ref. D4, 2392, (1971).

3. D. J. Gross, S. B. Treiman, Phys. Rev. D4, 1059, (1971).

4. M. Gell–Mann, Schladming Lectures 1972, CERN–preprint TH 1543.
W. A. Bardeen, H. Fritzsch, M. Gell–Mann, Proceedings of the Topical Meeting on Conformal Invariance in Hadron Physics, Frascati, May 1972.

5. J. Wess (Private communication to B. Zumino).

6. H. Fritzsch, M. Gell–Mann and A. Schwimmer, to be published.
D. J. Broadhurst and R. Jaffe, to be published.

7. H. Leutwyler, J. Stern, Nuclear Physics B20, 77 (1970).

8. R. Jackiw, DESY Summer School Lectures 1971, preprint MIT–CTP 236.

9. G. Domokos, S. Kövesi–Domokos, John Hopkins University preprint C00–3285–22, 1972.

10. A. Neveu, J. Scherk, Nuclear Physics B36, 155, 1972.

11. R. Dashen, M. Gell–Mann, Phys. Rev. Letters 17, 340 (1966).
M. Gell–Mann, Erice Lecture 1967, in: Hadrons and their Interactions, Academic Press, New York–London, 1968.
S.–J. Chang, R. Dashen, L. O'Raifeartaigh, Phys. Rev. 182, 1805 (1969).

12. P. Minkowski, unpublished (private communication).

Advantages of Color Octet Gluons

Harald Fritzsch

H. Fritzsch (✉)
Physik-Department, Ludwig-Maximilians-Universität Physik-Department, München, Germany
e-mail: fritzsch@mppmu.mpg.de

H. Fritzsch (ed.), *Murray Gell-Mann and the Physics of Quarks*, Classic Texts in the Sciences, https://doi.org/10.1007/978-3-319-92195-2_10

Volume 47B, number 4 PHYSICS LETTERS 26 November 1973

ADVANTAGES OF THE COLOR OCTET GLUON PICTURE☆

H. FRITZSCH*, M. GELL-MANN and H. LEUTWYLER**
California Institute of Technology, Pasadena, Calif. 91109, USA

Received 1 October 1973

It is pointed out that there are several advantages in abstracting properties of hadrons and their currents from a Yang–Mills gauge model based on colored quarks and color octet gluons.

In the discussion of hadrons, and especially of their electromagnetic and weak currents, a great deal of use has been made of a Lagrangian field theory model in which quark fields are coupled symmetrically to a neutral vector "gluon" field. Properties of the model are abstracted and assumed to be true for the real hadron system. In the last few years, theorists have abstracted not only properties true to each order of the coupling constant (such as the charge algebra $SU_3 \times SU_3$ and the manner in which its conservation is violated) but also properties that would be true to each order only if there were an effective cutoff in transverse momentum (for example, Bjorken scaling, V-A light cone algebra, extended V-A-S-T-P light cone algebra with finite quark bare masses, etc.).

We suppose that the hadron system can be described by a theory that resembles such a Lagrangian model. If we accept the stronger abstractions like exact asymptotic Bjorken scaling, we may have to assume that the propagation of gluons is somehow modified at high frequencies to give the transverse momentum cutoff. Likewise a modification at low frequencies may be necessary so as to confine the quarks and antiquarks permanently inside the hadrons.

The resulting picture could be equivalent to that emerging from the bootstrap-duality approach (in which quarks and gluons are not mentioned initially), provided the baryons and mesons then turn out to behave as if they were composed of quarks and gluons.

We assume here the validity of quark statistics (equivalent to para-Fermi statistics of rank three, but with restriction of baryons to fermions and mesons to bosons). The quarks come in three "colors", but all physical states and interactions are supposed to be singlets with respect to the SU_3 of color. Thus, we do not accept theories in which quarks are real, observable particles; nor do we allow any scheme in which the color non-singlet degrees of freedom can be excited. Color is a perfect symmetry. (We should mention that even if there is a fourth "charmed" quark u′ in addition to the usual u, d, and s, there are still three colors and the principal conclusions set forth here are unaffected.)

For a long time, the quark-gluon field theory model used for abstraction was the one with the Lagrangian density

$$L = -\bar{q}\,[\gamma_\alpha(\partial_\alpha - igB_\alpha\lambda_0) + M]\,q + L_B. \qquad (1)$$

Here M is the diagonal mechanical mass matrix of the quarks and L_B is the Lagrangian density of the free neutral vector field B_α, which is a color singlet. Recently, it has been suggested [1] that a different model be used, in which the neutral vector field $B_{A\alpha}$ is a color octet ($A = 1 \ldots 8$) and we have

$$L = -\bar{q}\,[\gamma_\alpha(\partial_\alpha - igB_{A\alpha}\chi_A) + M]\,q + L_B \text{ (Yang–Mills)}, \qquad (2)$$

where χ_A is the color SU_3 analog λ_i. In this communication we discuss the advantages of abstracting properties of hadrons from (2) rather than (1).

We remember, of course, that the real description of hadrons may involve a mysterious alteration of L_B to $\hat{L}_B$ or of L_B(Y-M) to $\hat{L}_B$(Y-M), where the new

☆ Work supported in part by the U.S. Atomic Energy Commission. Prepared under Contract AT(11-1)-68 for the San Francisco Operations Office, U.S. Atomic Energy Commission. Work supported in part by a grant from the Alfred P. Sloan Foundation.

* On leave from Max-Planck-Institut für Physik und Astrophysik, München, Germany.
** On leave from Institute for Theoretical Physics, Bern, Switzerland.

Volume 47B, number 4 PHYSICS LETTERS 26 November 1973

Lagrangian has the needed properties at high and low frequencies to give scaling and confinement respectively. No convincing example of such a situation has ever been given. In ref. [1], it was suggested the required new gluon propagation might be supplied in a model where $B_{A\mu}$ appears as one mode of a quantized string in a multilocal field theory version of a dual picture for the glue. (The mass-shell version of such a dual scheme, for particles treated as real, is known to reduce to a Yang–Mills theory as the slope parameter α' for Regge trajectories tends toward zero.) Another suggestion [2] is that somehow the free gluon propagator contains, instead of the factor $1/q^2$, a factor μ^2/q^4, where μ is some mass. All such suggestions are, for the moment, mere speculations.

It may be, of course, that there is no modification at high frequencies, in which case we would probably not have exact asymptotic Bjorken scaling. Also, modification at low frequencies may not be necessary for confinement.

A modified theory would clearly have an operator term δ in the energy density that violates scale invariance but not $SU_3 \times SU_3$, while the unmodified one would either lack δ or generate it spontaneously. A theory with $\delta = 0$ would have a massless scalar dilaton as $M \to 0$.

The simplest and most obvious advantage of (2) over (1) is that the gluons are now just as fictitious as the quarks. The color octet gluon field $B_{A\alpha}$ does not communicate with any physical channel, since the physical states are all color singlets; in contrast, the color singlet gluon field B_α would have the same quantum numbers as the baryon current, the ϕ meson, and so forth. Since in (2) the gluon is unphysical, we have no objection to the occurrence of long-range forces in its fictitious channel, produced either by massless gluons in the unmodified version or by the noncanonical glue propagation in the modified version. These fictitious long-range forces and the associated infrared divergencies could provide a mechanism for confining all color nonsinglets permanently. They would not be present in physical hadronic interactions, where long-range forces are know to be absent.

The second advantage is that we can see in (2) a *hint* as to why Nature selects color singlets. Looking at the crudest nonrelativistic, weak-coupling approximation to (2), we find a potential

$$g^2 (2\pi)^{-1} \sum_{i \neq j} r_{ij}^{-1} C_{iA} C_{jA} ,$$

where the C_{iA} are the color octet SU_3 charges of the various quarks, antiquarks, and gluons. Then it is easy to envisage a situation in which the only states with deep attraction would be the color singlets. (We suppose that in the true theory the other states become completely unphysical.)

Recently, this point has been given publicity by Lipkin [3], who treats, however, a Han–Nambu picture in which color nonsinglets can be physically excited by electromagnetism and in which there are three triplets of real quarks with integral charges that average to 2/3, −1/3, and −1/3. We have rejected such a picture. In fact, a serious argument against it is the clash between the color octet Yang–Mills gauge on the one hand and the electromagnetic gauge or the Yang–Mills gauge of unified weak and electromagnetic interactions on the other. Since, in our work, the weak and electromagnetic currents form color singlets, we encounter no such difficulty.

A third and very important advantage of the color octet gluon scheme has been pointed out by L.B. Okun in a private communication to H. Pagels. Okun's point is that in (1) there is no distinction between ordinary SU_3 and the SU_3 of color in the limit $m_u = m_d = m_s$, and thus we would have the symmetry of SU_9 (or of SU_6 for $m_u = m_d$) where these groups combine color SU_3 and ordinary SU_3. No evidence of such extended symmetries exists. In (2), of course, these annoying symmetries are not present.

A fourth apparent advantage of the color octet gluon scheme has recently been demonstrated [4] using the asymptotic perturbation theory method of Gell-Mann and Low. Assuming that the method is valid (sum of asymptotic forms of orders of perturbation theory equaling asymptotic form of sum), one can have a situation in which the bare coupling constant is zero, there are no anomalous dimensions for color singlet quantities, and the behavior of light cone commutators comes closer to scaling behavior than in the color singlet vector gluon case (1). However, actual Bjorken scaling does not occur; instead, each moment $\int F_2(\xi) \xi^n \, d\xi$ of the Bjorken scaling function appears multiplied, in the Bjorken limit, by a distinct power $(\ln q^2)^{p_n}$, where $-q^2$ is the virtual photon mass squared.

Volume 47B, number 4 PHYSICS LETTERS 26 November 1973

That sort of violation of Bjorken scaling is not contradicted by present experiments. Furthermore, many sum rules and symmetry principles of light cone current algebra would be preserved.

For us, the result that the color octet field theory model comes closer to asymptotic scaling than the color singlet model is interesting, but not necessarily conclusive, since we conjecture that there may be a modification at high frequencies that produces true asymptotic scaling.

There is one more advantage of the color octet gluon scheme over the color singlet scheme, and it is the main point we wish to stress in this communication. In either scheme, there is an anomalous divergence of the axial vector baryon current $F^5_{i\alpha}$. While, for the other eigth axial vector currents $F^5_{i\alpha}$ $(i = 1 \ldots 8)$, we have simply

$$\partial_\alpha F^5_{i\alpha}(x) = \mathcal{D}(x, x, \mathrm{i}\gamma_5\{\tfrac{1}{2}\lambda_i, M\}), \tag{3}$$

the divergence equation for $F^5_{0\alpha}$ is [5]

$$\partial_\alpha F^5_{0\alpha} = \mathcal{D}(x,x,\mathrm{i}\sqrt{\tfrac{2}{3}}M\gamma_5) + \sqrt{6}\,g^2\,(8\pi^2)^{-1}G_{\mu\nu}G^*_{\mu\nu}, \tag{4}$$

where $\mathcal{D}(x,y,G)$ is the physical operator that corresponds in a free quark model to $\bar{q}(x)Gq(y)$, and $G_{\mu\nu} = \partial_\mu B_\gamma - \partial_\nu B_\mu$ for the color singlet case, while $G_{A\mu\nu} = \partial_\mu B_{A\nu} - \partial_\nu B_{A\mu} + g f_{ABC} B_{B\mu} B_{C\nu}$ for the color octet case.

Here the extra term in (4) arises from a several-gluon effect in the strong interaction analogous to the two-photon effect in the familiar electromagnetic triangle anomaly [6], which contributes a term $e^2(16\pi^2)^{-1}F_{\mu\nu}F^*_{\mu\nu}$ to the divergence of F^5_3.

It was shown [6] that in renormalizable gluon models the anomalous divergence arises essentially from the lowest order triangle diagram.

Wilson has demonstrated [7] that the anomaly is the consequence of a singularity in coordinate space. In field theory models this singularity comes from low order quark loop diagrams, since higher order corrections are less singular and do not contribute. Therefore, in a theory in which the gluon propagation is less singular at small distances than in the canonical one, the anomaly coefficient will be unchanged, since the quark propagation is left canonical.

In the color singlet gluon picture, the anomalous divergence term in (4) is necessarily associated [5] with an anomalous singularity in the bilocal current $F^5_{0\alpha}(x,y)$ as $z^2 = (x-y)^2$ tends to zero:

$$F^5_{0\alpha}(x,y) \mathrel{\hat{=}} 3\,\mathrm{i}(2\pi^2)^{-1}g\,G^*_{\alpha\beta}\,z_\beta(z^2)^{-1}. \tag{5}$$

The existence of such a term, while not contradicted by experiment so far, would destroy the light cone algebra *as a system* since one of the bilocal currents arising from commutation of two physical currents would be infinite on the light cone. In any case, we have assumed that the full light cone algebra is correct or at most violated by powers of logarithms, and we therefore cannot tolerate the term (5).

In ref. [5], this situation was posed as a puzzle: how to get rid of the anomalous singularity in $F^5_{0\alpha}(x,y)$, while retaining the anomalous divergence term for $\partial_\alpha F^5_{0\alpha}(x)$ given by triangle diagram.

The color octet gluon scheme solves the puzzle. The anomalous divergence term in $\partial_\alpha F^5_{0\alpha}(x)$ is unchanged, except for replacing $G_{\mu\nu}G^*_{\mu\nu}$ by $G_{A\mu\nu}G^*_{A\mu\nu}$, but it is now associated with a singularity as $z^2 \to 0$ not in $F^5_{0\alpha}(x,y)$, but in a different formal quantity, the corresponding color octet operator, which we may call $F^5_{0A\alpha}(x,y)$:

$$F^5_{0A\alpha}(x,y) \mathrel{\hat{=}} 3\,\mathrm{i}(2\pi^2)^{-1}g\,G^*_{A\alpha\beta}\,z_\beta(z^2)^{-1}. \tag{6}$$

Since $F^5_{0A\alpha}(x,y)$ is not a physical operator, being a color octet, we can have no objection to its being singular on the light cone.

To summarize, then, the fifth advantage of the color octet gluon scheme is that we get rid of the unacceptable anomalous singularity (5) in $F^5_{0\alpha}(x,y)$.

Now we can believe and make use of the anomalous divergence term in (4). This term looks as if it could be very useful in connection with the PCAC idea. Let us assume that the strong form of PCAC is correct [8]. Formally, we mean by this that as the bare quark masses tend to zero and the generators of $SU_3 \times SU_3$ become conserved, the conservation occurs according to the Nambu–Goldstone pattern, with eight massless pseudoscalar mesons. Physically, we mean that the real world of hadrons is not terribly far from such a situation, and not far at all from a situation with $SU_2 \times SU_2$ conserved and three massless pions. The bare quark masses are such that $m_\mathrm{u} \approx m_\mathrm{d} \ll m_\mathrm{s}$ and the ratios $M^2_\pi : M^2_\mathrm{K} : M^2_\eta$ are not very different from $0:1:4/3$.

It has always been a great mystery why, if we abstract relations from a field theory model like (1) or (2), we do not have in the limit $M \to 0$ the conservation

Volume 47B, number 4 PHYSICS LETTERS 26 November 1973

of nine axial vector currents and the existence of nine massless pseudoscalar mesons. Turning on the quark bare masses, with $m_u \approx m_d \ll m_s$, we would have four nearly massless pseudoscalar mesons instead of three, in bad disagreement with observation. To put in another way, as m_u and m_d tend to zero, we would have $U_2 \times U_2$ conservation and four massless pseudoscalar mesons.

The mystery might appear to be resolved, since the anomalous term in (4) breaks the conservation of $F^5_{0\alpha}$ even in the limit $M \to 0$ and so in that limit it looks as if there need not be a ninth massless pseudoscalar meson ‡, and in the limit $m_u \to 0$, $m_d \to 0$ it looks as if there need not be a fourht one.

Unfortunately, the extra term in (4) is itself a divergence of another (non-gauge invariant) pseudovector, and thus as $M \to 0$ we still have the conservation of a modified axial vector baryon charge; we must still explain why this new ninth charge seems to correspond neither to a parity degeneracy of levels nor to a massless Nambu–Goldstone boson as $M \to 0$.

It is important to find the explanation ‡. Assuming that strong PCAC does not fail, we conjecture that the question is closely related to the question of whether there are modifications of Yang–Mills gluon propagation and, if so, what is the nature of those modifications.

Two of us (H.F. and M.G-M.) would like to thank S. Adler, W.A. Bardeen, R. Crewther, H. Pagels and A. Zee for useful conversations and the Aspen Center for Physics for making those conversations possible.

‡ In ref. [5], the authors, appalled at the ancmalous singularity that accompanied the anomalous divergence in the color singlet gluon case, discussed the possibility of somehow getting rid of the anomalous divergence and finding a different explanation of the absence of a ninth pseudoscalar meson as $M \to 0$. The alternative explanation tentatively offered was that F^5_0, cummuting with $SU_3 \times SU_3$, could vanish in the limit $M \to 0$ when applied to "single particle states" instead of giving either parity doubling or a ninth massless pseudoscalar meson. However, using the full group $(SU_6)_{W}$,currents of the light-like vector, axial vector, and tensor charges, we wee that F^5_0 fails to commute with the tensor charges T_{ix} and T_{iy}, and all matrix elements of those charges would have to vanish between "single particle states". The same is true of the modified F^5_0 that includes the effect of the anomalous divergence. It seems unlikely that all "single particle" matrix elements of T_{ix} and T_{iy} vanis as $M \to 0$.

References

[1] H. Fritzsch and M. Gell-Mann, Proc. XVI Intern. Conf. on High energy physics, Chicago, 1972, Vol. 2, p. 135.
[2] K. Kaufmann, private communication.
[3] H. Lipkin (Weizmann Institute) preprint 1973.
[4] H.D. Politzer (Harvard) preprint 1973; D. Gross and F. Wilczek (Princeton) preprint 1973.
[5] H. Fritzsch and M. Gell-Mann, Proc. Intern. Conf. on Duality and symmetry in hadron physics (Weizmann Science Press, 1971).
[6] J. Schwinger, Phys. Rev. 82 (1951) 664.
S.L. Adler, Phys. Rev. 177 (1969) 2426.
J.S. Bell and R. Jackiw, Nuovo Cimento 60A (1969) 47.
S.L. Adler and W.A. Bardeen, Phys. Rev. 182 (1969) 1517.
[7] K. Wilson, Phys. Rev. 179 (1969) 1499.
[8] The weight of evidence is now in favor of strong PCAC. See H. Fritzch, M. Gell-Mann and H. Leutwyler, in preparation.

Lectures on Quarks

Harald Fritzsch

H. Fritzsch (✉)
Physik-Department, Ludwig-Maximilians-Universität Physik-Department, München, Germany
e-mail: fritzsch@mppmu.mpg.de

H. Fritzsch (ed.), *Murray Gell-Mann and the Physics of Quarks*, Classic Texts in the Sciences, https://doi.org/10.1007/978-3-319-92195-2_11

Acta Physica Austriaca, Suppl. IX, 733–761 (1972)

QUARKS *

BY

M. GELL-MANN
CERN - Geneva+

In these lectures I want to speak about at least two interpretations of the concept of quarks for hadrons and the possible relations between them.

First I want to talk about quarks as "constituent quarks". These were used especially by G. Zweig (1964) who referred to them as aces. One has a sort of a simple model by which one gets elementary results about the low-lying bound and resonant states of mesons and baryons, and certain crude symmetry properties of these states, by saying that the hadrons act as if they were made up of subunits, the constituent quarks q. These quarks are arranged in an isotopic spin doublet u, d and an isotopic spin singlet s, which has the same charge as d and acts as if it had a slightly higher mass.

---

*Lecture given at XI. Internationale Universitätswochen für Kernphysik, Schladming, February 21 - March 4, 1972.

+On leave from CALTECH, Pasadena. John Simon Guggenheim Memorial Fellow.

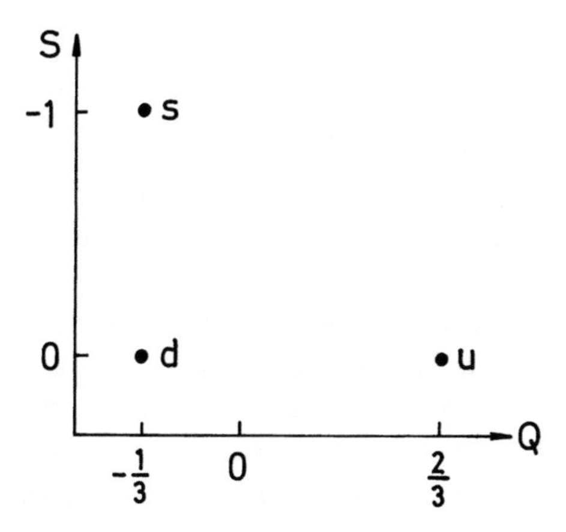

The antiquarks $\bar{q}$ of course, have the opposite behaviour. The low-lying bound and resonant states of baryons act like qqq and those of the mesons like $q\bar{q}$. Other configurations, e.g., $q\bar{q}q\bar{q}$, $qqqq\bar{q}$, etc., are called exotic, but they certainly exist in the continuum and may have resonances corresponding to them.

In this way one builds up the low-lying meson and baryon states and it is frequently useful to classify them in terms of an extremely crude symmetry group $U_6 \times U_6 \times O_3$, where one U_6 is for the quarks (three states of charge and two spin states) and one for the antiquarks, whereas O_3 represents a sort of angular momentum between them. This symmetry is however badly violated in the lack of degeneracy of the spectrum. The mesons then have as the lowest representation

$$(\underline{6}, \bar{\underline{6}}), \qquad L^P = 0^-$$

which gives the pseudoscalar and vector mesons, nine of each, just the ones which have been observed. (L=0 would normally have parity plus, but since we have a q and a $\bar{q}$

the intrinsic parity is minus.) The next pattern would be

$$(\underline{6}, \bar{\underline{6}}), \qquad L^P = 1^+$$

and this gives us the tensor mesons, axial vector mesons, another kind of axial vector mesons with opposite charge conjugation, and scalar mesons. All of these kinds have been seen, although not yet quite nine in every case. Then one can go up with L=2,3,... where there is just scattered experimental information. But whatever experimental information exists is at least compatible with this trivial picture. As is well known this group is not very well conserved in the spectrum and the states are badly split, e.g., m_π=140 MeV and $m_{\eta'}$=960 MeV.

For the baryons the lowest configuration is assigned to

$$(\underline{56}, \underline{1}), \qquad L^P = 0^+$$

that is three quarks in a totally symmetric state of spin, isospin, etc., without antiquarks; the parity is plus by definition. This gives the baryon octet with spin $\frac{1}{2}$ and just above it the decimet with spin $\frac{3}{2}$, which agree with the lowest-lying and best-known states of the baryons. The next thing one expects would be an excitation of one unit of angular momentum which changes the symmetry to the mixed symmetry under permutations:

$$(\underline{70}, \underline{1}), \qquad L^P = 1^-$$

and that seems to be a reasonable description of the low lying states of reversed parity. If one goes on to higher configurations things become more uncertain both experimentally and theoretically; presumably

$$(\underline{56}, \underline{1}), \qquad L^P = 2^+$$

exists and contains the Regge excitations of the corresponding ground state, likewise

$$(\underline{70}, \underline{1}), \; L^P = 3^- \qquad \text{and so on.}$$

In doing this we have to assume something peculiar about the statistics obeyed by these particles, if we want the model to be simple. One expects the ground state to be totally symmetric in space. If the quarks obeyed the usual Fermi-Dirac statistics for spin $\frac{1}{2}$ particles, then there would be an over-all antisymmetry and one would obtain a totally antisymmetric wave function in spin, isospin and strangeness, whereas ($\underline{56}$, $\underline{1}$) is the totally symmetric configuration in these quantum numbers. What most people have assumed therefore from the beginning (1963) is that the quarks obey some unusual kind of statistics in which every set of three has to be symmetrized but all other bonds have to be made antisymmetric, so that, e.g., two baryons are antisymmetric with respect to each other. One version of this came up under the name of parastatistics, precisely "para-Fermi statistics of rank three", which gives a generalization of the result I just described. I will discuss it in a slightly different way, which is equivalent to para-Fermi statistics of rank three with the restriction that physical baryon states are all fermions and physical meson states are all bosons.

We take three different kinds of quarks, that is nine altogether, and call the new variable distinguishing the sets "color", for example red, white and blue (R-W-B).

The nine kinds of quarks are then individually Fermi-Dirac particles, but we require that _all physical baryon and meson states be singlets under the SU_3 of "color"_. This means that for the meson $q\bar{q}$ configuration we now have

$$q_R\bar{q}_R + q_W\bar{q}_W + q_B\bar{q}_B$$

and for a baryon qqq we have

$$q_Rq_Wq_B - q_Wq_Rq_B + q_Bq_Rq_W - q_Rq_Bq_W + q_Wq_Bq_R - q_Bq_Wq_R \quad ,$$

which is totally antisymmetric in color and permits the baryon to be totally symmetric in the other variables space, spin, isospin and strangeness. This restriction to color singlet states for real physical situations gives back exactly the sort of statistics we want.

Now if this restriction is applied to all real baryons and mesons, then the quarks presumably cannot be real particles. Nowhere have I said up to now that quarks have to be real particles. There might be real quarks, but nowhere in the theoretical ideas that we are going to discuss is there any insistence that they be real. The whole idea is that hadrons act as if they are made up of quarks, but the quarks do not have to be real.

If we use the quark statistics described above, we see that it would be hard to make the quarks real, since the singlet restriction is not one that can be easily applied to real underlying objects; it is not one that factors: a singlet can be made up of two octets and these can be removed very far from each other such that the system over-all still is a singlet, but then we see the two pieces as octets because of the factoring property of the S matrix. If we adopt this point of view we are then faced with two alternatives: one is that there are three quarks, fictitious and obeying funny statistics; the other is that there are actually three triplets of real quarks, which is possible but unpleasant. In the latter case we would replace the singlet restriction with the assumption that the low lying states are singlets and one has to pay a large price in energy to get the colored SU_3 excited. I would prefer to adopt the first point of view, at least for these lectures.

Various crude symmetries and other related methods have been applied to these constituent quarks. First of all there is the famous subgroup of the classifying $U_6 \times U_6 \times U_3$, namely $[U_6]_W \times [O_2]_W$ which is applied to processes involving only one direction in space, like a vertex or forward and backward scattering (in general, collinear processes). $[O_2]_W$ has the generator L_z (assuming z is the chosen direction) and $[U_6]_W$ consists of the generators

$$\frac{1}{2}(\sum_i \lambda_i + \sum_j \lambda'_j),$$

$$\frac{1}{2}(\sum_i \lambda_i \sigma_{iz} + \sum_j \lambda'_j \sigma'_{jz}),$$

$$\frac{1}{2}(\sum_i \lambda_i \sigma_{ix} - \sum_j \lambda'_j \sigma'_{jx}),$$

$$\frac{1}{2}(\sum_i \lambda_i \sigma_{iy} - \sum_j \lambda'_j \sigma'_{jy}),$$

where the sum over i extends over the constituent quarks and the primed sum over j extends over the constituent antiquarks; we have 36 operations. There is a very crude symmetry of collinear processes under this group.

Another thing that has been done is to draw simple diagrams following quark lines through the vertices and the scattering. These have been recently used by Harari and Rosner, who called them "twig" diagrams after Zweig, who introduced them in 1964.

The twig diagram, e.g., for a meson-meson-meson vertex, looks like

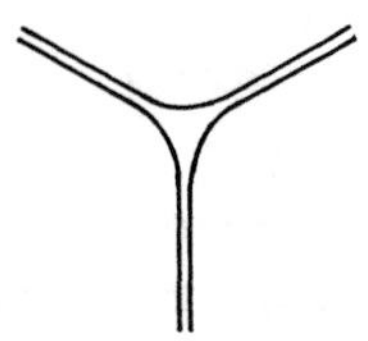

But another form

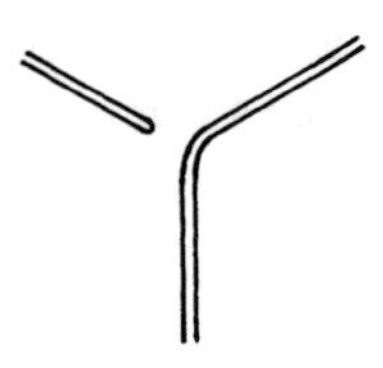

is forbidden by "Zweig's rule". This rule then leads to important experimental results, especially that the ϕ cannot decay appreciably into a ρ and π ($\phi \not\to \rho + \pi$), since the ϕ is composed of strange and antistrange quarks whereas ρ and π have only ordinary up and down quarks, and, therefore, the decay could take place only via the forbidden diagram. Similarly, we have the baryon-baryon-meson vertex

740

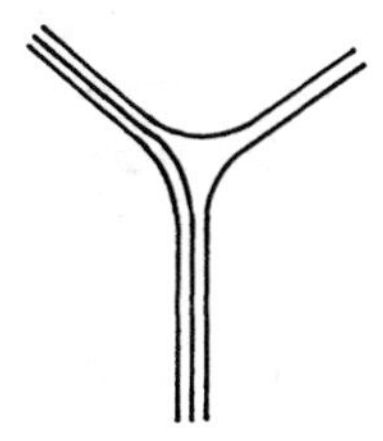

One can extend this concept to scattering processes and get a graphical picture of the so-called duality approach to scattering, e.g., for meson-meson scattering one can introduce the following diagram:

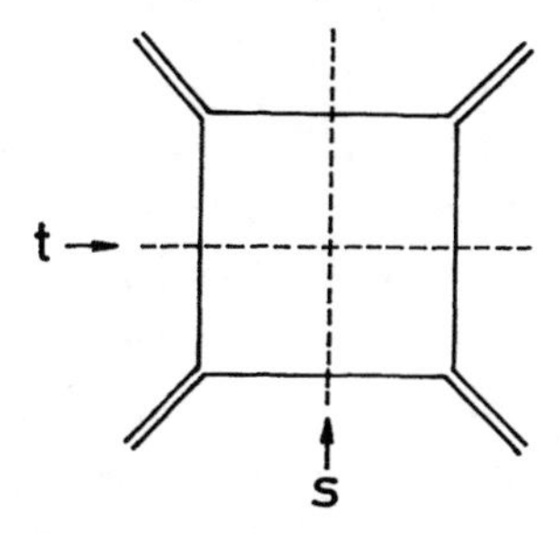

If we cut the diagram in the s and t channels we get $q\bar{q}$ in both cases: therefore, in meson-meson scattering we have ordinary non-exotic mesons in the intermediate states and exchange non-exotic mesons. We run into something of a trap, though, if we try to apply this to baryon-antibaryon scattering, because then we have a situation like

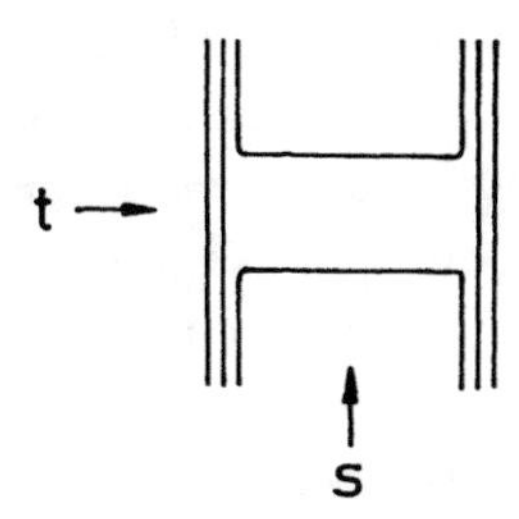

where the intermediate state is $q\bar{q}q\bar{q}$, which includes exotic configurations. In order to interpret this inconsistency different people have done different things.

The diagrams have been used in two different ways: one involves saying what the diagram means mathematically, and the other one involves not saying what it means mathematically. This is possible since here we do not have a priori a definite mathematical rule for computing the diagram, in contrast to a Feynman diagram for which a specific integral always exists. But we can obtain some results by never giving such a rule, only by noticing that we have zero when there is no diagram. Those so-called "null-relations" have been used by Schmid, by Harari and Rosner, by Zweig, Weyers and Mandula and by others for establishing a number of extremely useful sum rules. They give a correspondence between lack of resonances in the s channel, in places where the resonances would have to be exotic, and exchange degeneracy in the t channel. Exchange degeneracy is a noticeable feature of low energy hadron physics and a number of cases of agreement with experiments have been obtained.

All I want to say about the null-relation approach is that from the point of view of constituent quarks we are dealing here with a non-exotic approximation, because we are leaving out exotic exchanges, and that cannot be expected to be completely right. The simple null-equation duality approach is just another feature of the same kind of approximation we were talking about before, i.e., the classification under $U_6 \times U_6 \times O_3$ and the rough symmetry of collinear processes under $[U_6]_W \times [O_2]_W$, and when it fails that resembles a failure of such an approximation.

Another school of people consists of those who do the Veneziano duality kind of work and actually attempt to assign mathematical meaning to these diagrams. They go very far and construct almost complete theories of hadron scattering by means of extending these simple diagrams to ones with any number of quark pairs, but they run into trouble with negative probabilities or negative mass squares and the difficulty of introducing quark spin. There are also some difficulties with high energy diffraction scattering, etc. So that approach is not yet fully successful, while the much more modest null-relation approach has borne some fruit. However, if they overcome their difficulties, the members of the other school will have produced a full-blown hadron theory and advanced physics by a huge step.

There is a second use of quarks, as so-called "current quarks", which is quite different from their use as constituent quarks; we have to distinguish carefully between the two types in order to think about quarks in a reasonable manner. Unfortunately many authors including, I regret to say, me, have in past years written things that tended to confuse the two. In the following discussion of current quarks we attempt to write down properties that may be exact, at least to all orders in the strong interaction, with the weak, electromagnetic and gravitational interactions treated as perturbations. (It is necessary always to include gravity because the first order coupling to gravity is the stress-energy-momentum tensor and the integral over this tensor gives us the energy and momentum which we have to work with.)

When I say we attempt exact statements I do not mean that they are automatically true - there is also the incidental matter that they have to be confirmed by experiment, but the statements have a *chance* to be exact. Such statements which are supposed to be exact at least in certain limits or in certain well-defined approximations, or even generally exact, are to be contrasted with statements which are made in an ill-defined approximation or a special model whose domain of validity is not clearly specified. One frequently sees allegedly exact statements mixed up with these vague model statements and when experiments confirm or fail to confirm them it does not mean anything. Of course, we all have to work occasionally with these vague models because they give us some insight into the problem but we should carefully distinguish highly model-dependent statements from statements that have the possibility of being true either exactly or in a well defined limit.

The use of current quarks now is the following:we say that currents act as if they were bilinear forms in a relativistic quark field.We introduce a quark field,presumably one for the red,white and blue quarks and then we have for the vector currents in weak and electromagnetic interaction

$$F_{i\mu} \sim i\bar{q}_R \gamma_\mu \frac{\lambda_i}{2} q_R + i\bar{q}_W \gamma_\mu \frac{\lambda_i}{2} q_W + i\bar{q}_B \gamma_\mu \frac{\lambda_i}{2} q_B \quad .$$

The symbol $\sim$ means the vector current "acts like" this bilinear combination. Likewise the axial vector current acts like

$$F^5_{i\mu} \sim i\bar{q}_R \gamma_\mu\gamma_5 \frac{\lambda_i}{2} q_R + i\bar{q}_W \gamma_\mu\gamma_5 \frac{\lambda_i}{2} q_W + i\bar{q}_B \gamma_\mu\gamma_5 \frac{\lambda_i}{2} q_B \quad .$$

The reason why I want all these colors at this stage is that I would like to carry over the funny statistics for the current quarks, and eventually would like to suggest a transformation which takes one into the other, conserving the statistics but changing a lot of other things. An important feature of this discussion will be the following: is there any evidence for the current quarks that they obey the funny statistics? the answer is yes, and the evidence depends on a theoretical result due to many people but principally S.Adler.

The result is that in the PCAC limit one can compute exactly the decay rate of $\pi^o \to 2\gamma$. The basis on which Adler derived it was a relativistic renormalized quark-gluon field theory treated in renormalized perturbation theory order by order, and there the lowest order triangle diagram gives the only surviving result in the PCAC limit:

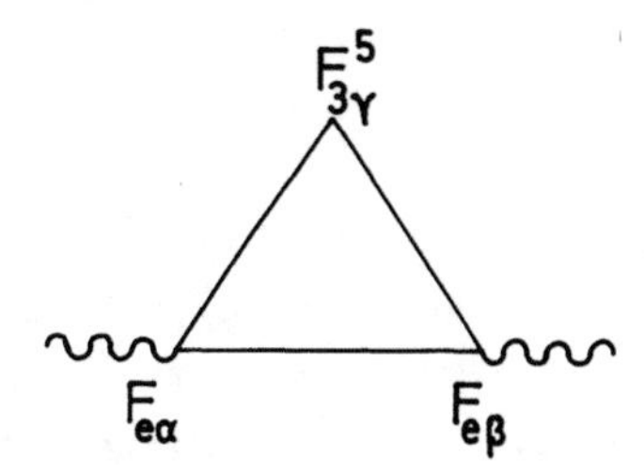

Here F_e are the electromagnetic currents, F^5_3 the third component of the axial current that is converted into a π^o through PCAC. We reject this derivation because order by order evaluation of a renormalizable quark-gluon field theory does not lead to scaling in the deep inelastic limit. Experiments at SLAC up to the present time are incapable of proving or disproving such a thing as scaling in the Bjorken limit but they are certainly suggestive and we would like to accept the Bjorken scaling. So we must reject the basis of Adler's derivation but we can derive this result in other ways, consistent with scaling, as I shall describe briefly later on.

What is sometimes said about Adler's computation is that this result completely contradicts the quark model. What is true is that it completely contradicts a hypothetical quark model that practically nobody wants, with Fermi-Dirac statistics. It agrees beautifully on the other hand with the ancient model that nobody would conceivably believe today in which things are made up of neutrons and protons. If you take basic neutrons and protons you get for an over-all coefficient of that diagram the following: we have charges squared multiplied by the third component of I, which means +1 for up isotopic spin and -1 down isotopic spin, so for protons and neutrons as basic constituents we get

$$(p, n): +1(1)^2 -1(0)^2 = 1$$

and with this Adler obtained exactly (within experimental errors) the right experimental decay rate for π^0 and even the right sign for the decay amplitude. If we take quarks u, d, s we get $(u, d, s): +1(2/3)^2-1(-1/3)^2=1/3$. So we obtain a decay rate which is wrong by a factor of 9. However, if we have the funny statistics - say in the easiest way with the red, white and blue color - we should put in

$$\begin{matrix} u_R, d_R, s_R \\ u_W, d_W, s_W \\ u_B, d_B, s_B \end{matrix} \quad : \quad 3[+1(2/3)^2-1(-1/3)^2] = 1 \quad ,$$

remembering that the current is a singlet in R-W-B, but in the summation we obtain a loop for each color. So we get back the correct result. Thus, there is this, in my mind, very convincing piece of evidence from the current quarks too for the funny statistics such as the constituent quarks seem to obey. The transformation between them should preserve statistics and so the picture seems to be a consistent one.

746

The relation "acts like" ($\sim$) which we used to define the current quarks can be strengthened as we introduce more and more properties of the currents which are supposed to be like the properties of these expressions. In other words there will be a hierarchy of strength of abstraction from such a field theory to the properties that we suggest are the exact characteristics of the vector and axial vector currents. We have to be very careful then to abstract as much as we can so as to learn as much as we can from the current quark picture, but not to abstract too much, otherwise first of all experiments may prove us wrong, and secondly that it may involve us with the existence of actual quarks, maybe even free quarks - and that, of course, would be a disaster.

If quarks are only fictitious there are certain defects and virtues. The main defect would be that we will never experimentally discover real ones and thus will never have a quarkonics industry. The virtue is that then there are no basic constituents for hadrons - hadrons act as if they were made up of quarks but no quarks exist - and, therefore, there is no reason for a distinction between the quark and bootstrap picture: they can be just two different descriptions of the same system, like wave mechanics and matrix mechanics. In one case you talk about the bootstrap and when you solve the equations you get something that looks like a quark picture; in the other case you start out with quarks and discover that the dynamics is given by bootstrap dynamics. An important question about the bootstrap approach is whether the bootstrap gives everything or whether there are some symmetry considerations to be supplied aside from the bootstrap. I do not know the answer, but some people claim to have direct information from heaven about it.

Let us go back to the current quarks. Besides the V and A currents we might have well defined tensor (T), scalar (S) and pseudoscalar (P) currents, which would act like

$$T_{i\mu\nu} \sim \bar{q}\,\lambda_i\,\sigma_{\mu\nu}\,q$$

$$S_i \sim \bar{q}\,\lambda_i\,q$$

$$P_i \sim i\bar{q}\,\lambda_i\,\gamma_5\,q \quad .$$

I think these currents all can be physically defined: the S and P currents would be related to the divergences of the V and A currents and the tensor currents would arise when you commute the currents with their divergences.

The first of the most elementary abstractions was the introduction of the $SU_3 \times SU_3$ charges, that is

$$\int F_{io}\, d^3x = F_i$$

$$\int F^5_{io}\, d^3x = F^5_i$$

with their equal time commutators. We do not have very good direct evidence that this is true, but the best evidence comes from the Adler-Weisberger relation which is in two forms: first the pure one, namely just these commutators

$$[F^5_i, F^5_j] = i\, f_{ijk}\, F_k$$

which give sum rules for neutrino reactions, and a second form which involves the use of PCAC giving sum rules for pion reactions, and those have been verified. So an optimist would say that the commutation relations are okay and PCAC is okay; the pessimist might say that they are both wrong but they compensate. That will be checked relatively soon as neutrino experiments get sophisticated enough to test the pure form. In the meantime I will assume that the equal-time commutators (ETC) and PCAC are okay. In order to get the Adler-Weisberger relation it is necessary to apply the ETC of light-like charges for the special kinematic condition of infinite momentum (in the z direction). We make the additional assumption that between finite mass states we can saturate the sum over intermediate states by finite mass states. In the language of dispersion theory this amounts to an assumption of unsubtractedness; in the language of light-cone theory it amounts to smoothness on the light cone. In this way Adler and Weisberger derived their simple sum rule.

We are considering here the space integrals of the time components of the V and A currents, but not those of the x or y components at $P_z=\infty$. We are restricting ourselves, in order to have saturation by finite mass intermediate states, to "good" components of the currents, those with finite matrix elements at $P_z=\infty$. These are $F_{io} \approx F_{iz}$ and $F^5_{io} \approx F^5_{iz}$ at $P_z=\infty$. The "bad" components F_{ix}, F_{iy}, F^5_{ix}, F^5_{iy} have matrix elements going like P_z^{-1} at infinite P_z. The components $F_{iz}-F_{io}$ and $F^5_{iz}-F^5_{io}$ have matrix elements going like P_z^{-2} and are "terrible".

One generalization that we can make of the algebra of $SU_3 \times SU_3$ charges is to introduce the tensor currents $T_{i\mu\nu}$. In the case of the tensor currents the good components are

$$T_{ixo} \cong T_{ixz}$$
$$T_{iyo} \cong T_{iyz}$$

from which we construct the charges

$$"T_{ix}" = \int T_{ixo}\, d^3x$$

$$"T_{iy}" = \int T_{iyo}\, d^3x$$

at $P_z=\infty$ and adjoin them to F_i and F_i^5. Thus we get a system of 36 charges and that just gives us a $[U_6]_W$. One reason why I introduced these tensor currents is that it is simpler to work with $[U_6]_W$, which we have met before, rather than with one of its subgroups $SU_3\times SU_3$. These charges then form generators of an algebra which we call

$$[U_6]_{W,\infty,\text{currents}}\ ;$$

to be contrasted with

$$[U_6]_{W,\infty,\text{strong}}$$

which had to do with the constituent quarks. The contrast is between the $[U_6]_{W,\infty,\text{strong}}$ which is essentially *approximate* in its applications (collinear processes), and the $[U_6]_{W,\infty,\text{currents}}$ drawn from presumably exact commutation rules of the currents and having to do with current quarks. Although they are isomorphic they are not equal.

For those who cannot stand the idea of introducing the tensor currents, we can just reduce both groups to their subgroups $SU_3\times SU_3$. In that case for $[U_6]_{W,\infty,\text{currents}}$ we are discussing only the vector and axial vector charges and for $[U_6]_{W,\infty,\text{strong}}$ only the so-called coplanar subgroup. Then we can make the same remark that these two are not equal, they are mathematically similar but their matrix elements are totally different. So one of them is the transform of the other in some sense.

The transformation between current and constituent quarks is then phrased here in a way which does not involve quarks; we discuss it as a transformation between $[U_6]_{w,\infty,\text{currents}}$ and $[U_6]_{w,\infty,\text{strong}}$ (or their respective subgroups). What would happen if they were equal? We know that for $[U_6]_{w,\infty,\text{strong}}$ the low-lying baryon and meson states belong approximately to simple irreducible representations 35, 1, 56, etc. If this were true also of $[U_6]_{w,\infty,\text{currents}}$ then we would have the following results:

$$-\frac{G_A}{G_V} \cong \frac{5}{3}$$

which we know is more like $5/3\sqrt{2}$; the anomalous magnetic moments of neutron and proton would be approximately zero, while they are certainly far from zero, and so on.

Many authors have in fact investigated the mixing under this group, and they found that there is an enormous amount of mixing, e.g., the baryon is partly 56, $L_z=0$ and partly 70, $L_z=\pm 1$ and the admixture is of the order of 50%. There are higher configurations, too.

So these two algebras are not closely equal although they have the same algebraic structure. And there is some sort of a relation between them, which might be a unitary one, but we cannot prove that since they do not cover a complete set of quantum numbers. But we can certainly look for a unitary transformation connecting the two algebras and my student J. Melosh is pursuing that problem. He has found this transformation for free quarks where it is simple and leads to a conserved $[U_6]_{w,\text{strong}}$. But, of course, we are not dealing with free quarks and have to look at a more complicated situation. What I want to emphasize is that here we have the definition of the search; the search is for a transformation connecting the two algebras. In popular language we can refer to it as a search for the transformation connecting constituent quarks and current quarks.[+]

Let me mention here the work of another student of mine, Ken Young, who has cleaned up this past year at Caltech work that Dashen and I began about 7 years ago, and which we continued sporadically ever since. That is the attempt to represent the charges and also the transverse Fourier components of the charge densities at infinite momentum completely with non-exotic states, so as to make a non-exotic relativistic quark model as a representation of charge density algebra. We ran into all kinds of troubles, particularly with the existence of states with negative mass squares and the failure of the operators of different quarks to commute with each other. Young seems to have shown that these difficulties are a property of trying to represent the charge algebra at infinite momentum with non-exotics alone. Therefore, the transformation which connects the two algebras does not just mix up non-exotic states but also brings in higher representations that contain exotics. In simple lay language the transformation must bring in quark pair contributions and the constituent quark looks then like a current quark dressed up with current quark pairs.

We therefore must reject all the extensive literature, which I am proud not to have contributed to over the last few years, in which the constituent quarks are treated as current quarks and the electromagnetic current is made to interact through a simple current operator $F_{e\mu}$ with what

---

[+]Buccella, Kleinert and Savoy have suggested a simple phenomenological form of such a transformation.

are essentially constituent quarks. That is certainly wrong.

Another way of describing the infinite momentum and the smoothness assumption is to perform an alibi instead of an alias transformation, i.e., instead of letting everything go by you at infinite momentum you leave it alone at finite momentum and you run by it. These two are practically equivalent. In that case one is not talking about infinite momentum but about the behaviour in co-ordinate space as we go to a light-like plane and about the commutators of light-like charges which are integrated over a light-like plane instead of an equal-time plane. Leutwyler, Stern and a number of other people have especially emphasized this approach. From that again one can get the Adler-Weisberger relation, and so forth.

On the light-like plane (say z+t=0) we have the commutation rules not only for the charges, but also for the local densities of the good components of the currents, namely $F_{io}+F_{iz}$ and $F^5_{io}+F^5_{iz}$ for V and A, with the possible adjunction of the good components $T_{ixo}+T_{ixz}$ and $T_{iyo}+T_{iyz}$ of the tensor currents. Especially useful is the algebra of these quantities integrated over the variable z-t and Fourier-transformed with respect to the variables x and y. We obtain the operators $F_i(\vec{k}_\perp)$, $F^5_i(\vec{k}_\perp)$, $T_{ix}(\vec{k}_\perp)$, and $T_{iy}(\vec{k}_\perp)$ and they have commutation relations like

$$[F_i(\vec{k}_\perp), F_j(\vec{k}'_\perp)] = i\, f_{ijk}\, F_k(\vec{k}_\perp+\vec{k}'_\perp) \quad .$$

By the way, if we take this equation to first order in $\vec{k}_\perp$ and $\vec{k}'_\perp$ we get the so-called Cabibbo-Radicati sum rule for photon-nucleon collisions, which seems to work quite well.[+]

[+]Strictly, the operators $F_i(\vec{k}_\perp)$ are not exactly the Fourier transforms of integrals of current densities but rather these Fourier transforms multiplied by

$$\exp\{i[k_x(\Lambda_x+J_y)+k_y(\Lambda_y-J_x)][P_o+P_z]^{-1}\} \ ,$$

So far we have abstracted from quark field theory relations that were true not only in a free quark field model but also to all orders of strong interactions in a field model with interactions, say through a neutral "gluon" field coupled to the quarks. (We stick to a vector "gluon" picture so that we can use the scalar and pseudoscalar densities S_i and P_i to describe the divergences of the vector and axial vector currents respectively.) We shall continue to make use of abstractions limited in this way, since if we took all relations true in a free quark model we would soon be in trouble: we would be predicting free quarks! However, in what follows we consider the abstraction of propositions true only formally in the vector "gluon" model with interactions, not order by order in renormalized perturbation theory. We do this in order to get Bjorken scaling, which fails to each order of renormalized perturbation theory in a barely renormalizable model like the quark-vector "gluon" model but which, as mentioned before, we would like to assume true.

That leads us to light-cone current algebra, on which I have worked together with Harald Fritzsch. It has been studied by many other people as well, including Brandt and Preparata, Leutwyler et al., Stern et al., Frishman, and, of course, Wilson, who pioneered in this field although he disagrees with what we do nowadays. Those whose work is most similar to ours are Cornwall and Jackiw and Llewellyn-Smith.

The first assumption in light-cone current algebra is to abstract from free quark theory or from formal vector "gluon" theory the leading singularity on the light cone $(x-y)^2 \approx 0$ of the connected part of the commutator of two currents at space-time points x and y. For V and A currents we find

---

where $\vec{\Lambda}$ is the Lorentz-boost operator, $\vec{J}$ is the total angular momentum, and the component P_o+P_z of the energy-momentum-four-vector is conserved by all the operators $F_i(\vec{k}_\perp)$.

754

$$[F_{i\mu}(x),\ F_{j\nu}(y)] \hat{=} [F^5_{i\mu}(x),\ F^5_{j\nu}(y)] \hat{=}$$

$$\frac{1}{4\pi}\,\partial_\rho\{\varepsilon(x_o-y_o)\,\delta((x-y)^2)\}\{(i\ f_{ijk}-d_{ijk})(s_{\mu\nu\rho\sigma}F_{k\sigma}(y,x) + i\varepsilon_{\mu\nu\rho\sigma}F^5_{k\sigma}(y,x)) +$$

$$+ (i\ f_{ijk}+d_{ijk})(s_{\mu\nu\rho\sigma}F_{k\sigma}(x,y)-i\ \varepsilon_{\mu\nu\rho\sigma}F^5_{k\sigma}(x,y))\}\ ,$$

$$[F_{i\mu}(x),F^5_{j\nu}(y)] \hat{=}$$

$$\frac{1}{4\pi}\,\partial_\rho\{\varepsilon(x_o-y_o)\,\delta((x-y)^2)\}\{(i\ f_{ijk}-d_{ijk})(s_{\mu\nu\rho\sigma}F^5_{k\sigma}(y,x) + i\varepsilon_{\mu\nu\rho\sigma}F_{k\sigma}(y,x)) +$$

$$+ (i\ f_{ijk}+d_{ijk})(s_{\mu\nu\rho\sigma}F^5_{k\sigma}(x,y)-i\ \varepsilon_{\mu\nu\rho\sigma}F_{k\sigma}(x,y))\}\ .$$

On the right-hand side we have the connected parts of bilocal operators $F_{k\sigma}(x,y)$ and $F^5_{k\sigma}(x,y)$ that reduce to the local currents $F_{k\sigma}(x)$ and $F^5_{k\sigma}(x)$ as $y \to x$. The bilocal operators are defined only in the vicinity of $(x-y)^2=0$. Here $s_{\mu\nu\rho\sigma}=\delta_{\mu\rho}\delta_{\nu\sigma}+\delta_{\nu\rho}\delta_{\mu\sigma}-\delta_{\mu\nu}\delta_{\rho\sigma}$.

The formulae give Bjorken scaling by virtue of the finite matrix elements assumed for $F_{k\sigma}(x,y)$ and $F^5_{k\sigma}(x,y)$; in fact the Fourier transform of the matrix element of $F_{k\sigma}(x,y)$ is just the Bjorken form factor. The fact that all charged fields in the model have spin 1/2 determines the algebraic structure of the formula and gives the prediction $\sigma_L/\sigma_T \underset{Bj}{\rightarrow} 0$ for deep inelastic electron scattering, not in contradiction with experiment. The electrical and weak charges of the quarks in the model determine the coefficients in this formula, and give rise to numerous sum rules and inequalities for the SLAC-MIT experiments and for corresponding neutrino and antineutrino experiments in the Bjorken limit, none in contradiction with experiment, although the inequality $1/4 \leq F^{en}(\xi)/F^{ep}(\xi) \leq 4$ appears to be tested fairly severely at the lower end near $\xi=1$.

The formula for the leading light-cone singularity in the commutator contains, of course, the physical information that near the light-cone we have full symmetry with respect to $SU_3 \times SU_3$ and with respect to scale transformations in co-ordinate space. Thus there is conservation of dimension in the formula, with each current having $\ell=-3$ and the singular function of x-y also having $\ell=-3$.

A simple generalization of the abstraction we have considered turns it into a closed system, called the basic light-cone algebra. Here we commute the bilocal operators as well, for instance $F_{i\mu}(x,u)$ with $F_{j\nu}(y,v)$, as all the six intervals among the four space-time points approach 0, so that all four points tend to lie on a light-like straight line in Minkowski space. Abstraction from the model gives us on the right-hand side a singular function of one coordinate difference, say x-v, times a bilocal current $F_{k\sigma}$ or $F^5_{k\sigma}$ at the other two points, say y and u, plus an expression with (x,v) and (y,u) interchanged, and the system closes algebraically. The formulae are just like the ones for local currents.

We shall assume here the validity of the basic lightcone algebraic system, and discuss possible applications and generalizations.

First of all, we may consider what happens when the points x and u lie on a light-like plane with one value of z+t and y and v lie on another light-like plane with a slightly different value of z+t, and we let these values approach each other.

For commutators of good components of currents, the limit is finite, and we get a generalization of the light-plane algebra of commutators of good densities $F_{io}+F_{iz}$ and $F^5_{io}+F^5_{iz}$. There is now a fourth argument in each density, namely the internal co-ordinate η, which runs only in the light-like direction z-t. As before, we get the most useful results by integrating over the average z-t and Fourier-transforming with respect to the transverse average co-ordinates x and y, obtaining operators $F_i(\vec{k}_\perp, \eta)$, $F^5_i(\vec{k}_\perp, \eta)$, with commutation relations like

$$[F_i(\vec{k}_\perp, \eta), F_j(\vec{k}'_\perp, \eta')] = i f_{ijk} F_k(\vec{k}_\perp + \vec{k}'_\perp, \eta + \eta') \quad .$$

Remember that $\vec{k}_\perp$ is in momentum space and η in (relative) co-ordinate space.

The non-local operators $F_i(\vec{k}, \eta)$ acting on the vacuum create strings of mesons with all values of the meson spin angular momentum J. In fact, a power series expansion in η of $F_i(\vec{k}_\perp, \eta)$ is just an expansion in η^{J-1}. At large η, we can reggeize and obtain a dominant term in $\eta^{\alpha(-k^2_\perp)-1}$, where $\alpha(-k^2_\perp)$ is the leading Regge trajectory in the relevant meson channel, for instance P or ρ. (In the work on these questions, Fritzsch has played a particularly important role.) The couplings of the Regge poles in the bilocals are proportional to the hadronic couplings of meson Regge poles.

If we commute bad components with bad components as the two light planes approach each other, then the leading singularity on the light-cone leads to a singular term that goes like a δ-function of the difference in coordinates z+t. This singular term, which is multiplied by a good component of a bilocal current on the right-hand side, gives the Bjorken scaling in deep inelastic scattering. Unlike the good-good commutators on the light plane, it involves a commutator of local quantities on the left giving a bilocal on the right, a bilocal of which the matrix elements give the Fourier transforms of the Bjorken scaling functions $F(\xi)$.

We may now generalize, if we keep abstracting from the vector gluon model, to a connected light-cone algebra involving V, S, T, A and P densities, where the divergences of the V and A currents are proportional to the S and P currents respectively, with coefficients that correspond in the model to the three bare quark masses, forming a diagonal 3×3 matrix M. The divergences of the axial vector currents, for example, are given by masses M multiplied by normalized pseudoscalar densities P_i.

The scalar and pseudoscalar densities are all bad, and do not contribute to the good-good algebra on the light plane, but we can commute two of these densities as the light planes approach each other and obtain the singular Bjorken term. In fact, the leading light-cone singularity in the commutator of two pseudoscalars or two scalars just involves the same vector bilocal densities as the leading singularity in the commutator of two vector densities

$$[P_i(x), P_j(y)] \hat{=} [S_i(x), S_j(y)] \hat{=} [F_{i\mu}(x), F_{j\mu}(y)]$$

so that the P's and S's give Bjorken functions that are not only finite but known from deep inelastic electron and neutrino experiments. The Bjorken limit of the commutator of two divergences of vector or axial vector currents is also measurable in deep inelastic neutrino experiments, albeit very difficult ones, since they involve polarization and also involve amplitudes that vanish when the lepton masses vanish. The important thing is that the shapes of the form factors in such experiments are predictable from known Bjorken functions and the overall strength is given by the "bare quark mass" matrix M, which is thus perfectly measurable, according to our ideas, even though the quarks themselves are presumably fictitious and have no real masses.

The next generalization we may consider is to abstract the behaviour of current products as well as commutators near the light-cone. Here we need only abstract the principle that scale invariance near the light-cone applies to products as well as commutators. The result is that products of operators, and even physical ordered products, are given, apart from subtraction terms that act like four-dimensional δ functions, by the same expressions as commutators, with $\varepsilon(x_o-y_o)\,\delta((x-y)^2)$ replaced by $1/[(x-y)^2-i(x_o-y_o)\varepsilon]$ for ordinary products and by $1/[(x-y)^2-i\varepsilon]$ for physical ordered products. The subtraction terms can often be determined from current conservation; sometimes they are zero and sometimes, for certain processes, they do not matter even when they are non-zero.

Using the current products, one can design experiments to test the bilocal-bilocal commutators, for example fourth order cross-sections like those for $e^-+p\rightarrow e^-+X+\mu^++\mu^-$, where X is any hadronic state, summed over X.

Using products, and employing consistency arguments, we can determine the form of the leading light-cone singularity in the disconnected part of the current commutator, i.e., the vacuum expected value of a current commutator, and it turns out to be the same as in free quark theory or formal quark "gluon" theory. The constant in front is not determined in this way, and we must abstract it from the model. It depends on the statistics. With our funny "quark statistics" or with nine real quarks, the constant is three times as large as for three Fermi-Dirac quarks.

We can then predict the asymptotic cross-section for $e^+ + e^- \to$ hadrons using single photon annihilation, namely

$$\frac{\sigma(e^+ + e^- \to \text{hadrons})}{\sigma(e^+ + e^- \to \mu^+ + \mu^-)} \to 3[(\tfrac{2}{3})^2 + (-\tfrac{1}{3})^2 + (-\tfrac{1}{3})^2] = 2$$

where we would have obtained 2/3 with three Fermi-Dirac quarks.

We are now in a position to go back and rederive the Adler result for the rate of $\pi^0 \to 2\gamma$ in the PCAC approximation. Following the lead of Crewther, who first showed how such an alternative derivation could be given, Bardeen, Fritzsch and I use the connected light-cone algebra and the disconnected result just given to obtain the Adler result without invoking renormalized perturbation theory. The answer, as we indicated earlier, agrees with the experimental $\pi^0 \to 2\gamma$ amplitude in both sign and magnitude.

A final generalization, about which Fritzsch and I are not so convinced as we are of the others, involves a change in our approach from considering only quantities based on currents that couple to electromagnetism and the weak interaction to including quantities that are not physically determinable in that way. I have mentioned that bilocals like $F_{i\mu}(x,y)$, which are analogous to quantities in the model that involve one quark operator and one anti-quark operator, can be applied to the vacuum to create Regge sequences of non-exotic meson states. It might also be useful to define trilocals $B_{\alpha\beta\gamma,abc}(x,y,z)$ that are analogous to operators in the model involving three quark operators at x, y and z, when these points lie on the same straight light-like line, and to abstract their algebraic properties from the model, so that sequences of baryon states could be produced from the vacuum. We could, in fact, construct operators that would, between the vacuum and hadron states, give a partial Fock space for hadrons with any number of quarks and antiquarks lying on a straight light-like line. Whether this makes sense, and how many properties of hadrons we can calculate from such a partial Fock space of "wave functions" we do not know.

If we go too far in this direction, and try to construct a complete Fock space for quarks and antiquarks on a light-like plane, abstracting the algebraic properties from free quark theory, we are in danger of ending up with real quarks, and perhaps even with free real quarks, as mentioned before. In our work, we are always between Scylla and Charybdis; we may fail to abstract enough, and miss important physics, or we may abstract too much and end up with fictitious objects in our models turning into real monsters that devour us.

In connection with the written version of these lectures, I should like to thank Dr. Heimo Latal and his collaborators for the excellent lecture notes that they provided me. I should also like to thank Dr. Oscar Koralnik of Geneva for providing the beautiful table on which most of my writing was done. I acknowledge with thanks the hospitality of the Theoretical Study Division of CERN.

Printed in the United States
By Bookmasters